Lecture Notes in Computer Science 2046

Edited by G. Goos, J. Hartmanis and J. van Leeuwen

Springer

Berlin
Heidelberg
New York
Barcelona
Hong Kong
London
Milan
Paris
Singapore
Tokyo

Hans Daduna

Queueing Networks
with Discrete Time Scale

Explicit Expressions for the Steady State Behavior
of Discrete Time Stochastic Networks

 Springer

Series Editors

Gerhard Goos, Karlsruhe University, Germany
Juris Hartmanis, Cornell University, NY, USA
Jan van Leeuwen, Utrecht University, The Netherlands

Author

Hans Daduna
Universität Hamburg, Fachbereich Mathematik
Bundesstraße 55, 20146 Hamburg, Germany
E-mail: daduna@math.uni-hamburg.de

Cataloging-in-Publication Data applied for

Die Deutsche Bibliothek - CIP-Einheitsaufnahme

Daduna, Hans: Queueing networks with discrete time scale : explicit expressions for
the
steady state behavior of discrete time stochastic networks / Hans Daduna. -
Berlin ; Heidelberg ; New York ; Barcelona ; Hong Kong ; London ; Milan ;
Paris ; Singapore ; Tokyo : Springer, 2001
   (Lecture notes in computer science ; Vol. 2046)
   ISBN 3-540-42357-5

CR Subject Classification (1998): C.2, C.4, B.8, D.4.8, D.4.4, F.1, F.2.2, G.3

ISSN 0302-9743
ISBN 3-540-42357-5 Springer-Verlag Berlin Heidelberg New York

Springer-Verlag Berlin Heidelberg New York
a member of BertelsmannSpringer Science+Business Media GmbH

http://www.springer.de

© Springer-Verlag Berlin Heidelberg 2001
Printed in Germany

Typesetting: Camera-ready by author, data conversion by PTP-Berlin, Stefan Sossna
Printed on acid-free paper   SPIN: 10781454   06/3142   5 4 3 2 1 0

Preface

Networks of queues (= stochastic networks) have been a field of intensive research over the last three decades. The foundation for this research is classical queueing theory, which mainly dealt with single node queueing systems [21], [24], [49]. There is now a well developed theory of stochastic networks accompanied by an unbounded set of open problems which originates directly from applications as well as from theoretical considerations. Most of these open problems are easily stated and put into a theoretical framework, but often require either intricate techniques on an adhoc basis or deep mathematical methods. Often both of these approaches have to be combined to successfully tackle the solution of a quickly formulated problem.

The possibly most influential sources were the books by Kelly [75], Whittle [139], and Walrand [136]. The recent books by Van Dijk [40], Serfozo [121], and Chao, Miyazawa, Pinedo [18] gave a good state-of-the-art survey of the subject.

The development of queueing network theory which provided application areas with solutions, formulas, and algorithms commenced around 1950 in the area of Operations Research, with special emphasis on production, inventory, and transportation. The first breakthrough came with the works of Jackson [69] and Gordon and Newell [50]. Both papers appeared in OPERATIONS RESEARCH. The second breakthrough in queueing network theory was already connected with Computer Science: the celebrated papers by Baskett, Chandy, Muntz, and Palacios [4], which appeared in the JOURNAL OF THE ASSOCIATION FOR COMPUTING MACHINERY, and by Kelly [74].
At the same time the two volumes of Kleinrock's book appeared [81], [82]. From that time on queueing network theory and its application has been intimately connected with performance analysis of complex systems in Computer and Communication Sciences, both on the hardware and on the software level.

The way that production, manufacturing, and transportation is growing together with information processing and communication technology is resulting in more and more complex systems which require even more elaborate models, techniques, and algorithms to better understand their performance behavior and to predict performance and quality of service.

The classical single node queues and their networks are models living on a continuous time scale. Even if for some applications a discrete time scale might be more appropriate, the well established continuous time machinery often serves as an approximation tool. Consequently, the survey articles [24], [137] from 1990 still do not review discrete time models.

However, from then on an astonishing evolution of discrete time stochastic network models can be observed. Usually it is thought that the invention of ATM (Asynchronous Transfer Mode) as the protocol for high speed transmission network technology triggered this development.

At least three (to my knowledge) books appeared recently solely dedicated to theory and applications of discrete time queueing systems and networks, [14], [128], and [142]. The contents of this lecture notes are spread over the same area, but are almost complementary to the above mentioned books. The center of my interest is the question, whether there is in discrete time an analogue to the celebrated *product form calculus* of continuous time stochastic network theory. This calculus lies behind the two breakthroughs mentioned above. Its simplicity and general applicability was its entrance ticket into applications. Its theoretical elegance opened the door to mathematical investigations around stochastic networks.

From a mathematical point of view, for the description of stochastic networks, we deal with stochastic processes in continuous time as well as in discrete time. The theory of stochastic processes in continuous time is much more elaborate than that in discrete time. However, we shall see that we have to pay for turning to the simpler theoretical framework by being burdened with a much more technically involved machinery for arriving at simple methods for performance analysis and simple algorithms. The question of whether there can be a product form calculus in discrete time will be answered moderately positively in the book, but in my opinion the answer is still in a premature status. There seems to be much need for further research and there *are disappointing negative partial answers* in cases where the continuous time counterparts of the question have an easy positive answer. This will be discussed at several instances throughout the presentation.

The text is centered around explicit expressions for the steady state behavior of discrete time queueing networks. Three classes of networks which show a product form equilibrium will be discussed in detail:

(1) Linear networks (closed cycles and open tandems) of single server FCFS Bernoulli nodes.

(2) Networks of doubly stochastic and geometrical queues (which are discrete time analogues of Kelly's symmetric and general servers and of the BCMP nodes):
Customers of different types move through the network governed by a general routing mechanism. Request for service at symmetric servers may be according to general distributions.

(3) Networks with batch movements of customers and batch service. The service and routing mechanism is defined on the basis of an abstract transition scheme.

I further discuss computational algorithms for the standard performance measures and present explicit results on end–to–end–delay distributions and their moments.

The mathematical prerequisites for reading the book are moderate: discrete time Markov chains and their elementary linear algebra, and elementary transform methods (generating functions, z–transform) as can be found in [70], chapter 2, 3, and [71], chapter 10, 11, or [44], chapter 15, or [66], [67]. Knowledge of elementary continuous time queueing networks, [75], [47] would be helpful to understand their discrete time counterparts, but is not necessary.

I tried to compile the relevant literature with respect to discrete time stochastic networks. But, unfortunately enough, due to the enormous increase of publication rates in the field I am quite sure that I have missed some sources. However, I did not try to repeat the compilation in [18] on batch movement networks (with e.g. signals and further specific properties) where in general it is proposed, that the derivations and results hold for discrete time and continuous time systems equally. A number of models, which in my opinion are prototype models, are surveyed in chapter 6 with reference to further readings.

These lecture notes are based on the lecture notes of a one-semester course on "Discrete Time Stochastic Networks" held at the Department of Mathematics at Hamburg University in summer 1997 for students on a graduate level. The audience consisted of members of the Departments of Mathematics and Computer Sciences and of the Department of Communication Networks of the Technical University of Hamburg–Harburg. A preliminary version of these lecture notes was used as material and distributed within a Tutorial on "Discrete Time Queueing Networks: Recent Developments", which I held at the conference PERFORMANCE '96 in Lausanne (Switzerland) in October 1996.

The text is a report on some parts of the recent research and development in stochastic network theory and some of its applications. It may also serve as a textbook for a course on discrete time stochastic networks or as a companion book in a course on (elementary) stochastic processes.

Acknowledgements. I would like to thank students and colleagues who participated in the lectures on "Discrete Time Stochastic Networks" in summer 1997 for their many valuable comments and remarks on the problems I presented to them. I thank Kristin Betancourt for reading parts of the manuscript. I am grateful to Regina, Mirok Korea, and their friends for their encouragement and continuous support during the time I was writing these lecture notes.

March 2001   Hans Daduna

Contents

Chapter 1

Introduction

1.1 Introduction

The standard models of classical queueing theory are systems operating in continuous time. Applying such queueing models to predict the performance of
transmission channels dates back to the early times of Erlang around 1925 (see
[12]). Generalizations of his famous *Erlang's formulas for loss and waiting probabilities* are still of practical value in performance analysis. Even more: With
the advent of modern high speed transmission networks much more elaborated
techniques are developed to predict for these networks such probabilities which
count as important characteristics for their quality of service.

In the early times of queueing theoretical applications in telecommunication systems networks of transmission lines were modeled using brute force decomposition approximations – which are still in use and are considered as unavoidable in
many situations. Short descriptions are [47], chapter 5 for the continuous time
framework, [14], section 4.1.6 for discrete time networks.

First steps to overcome the restriction to decomposition approximations were
done around 1950 by coupling classical exponential queues in lines. These were
used to model jobshop like production systems, transportation lines, and complex distribution lines with inventories. Next step was the first breakthrough by
the works of Jackson [69] and Gordon and Newell [50] as part of Operations Research. In these networks indistinguishable customer traveled through nodes on
a network graph, who experienced delay at the service stations due to congestion
of the nodes, which originates from service requests of other customers.

For computing systems and their networks Kleinrock popularized queueing
network theory in continuous time as a powerful tool in performance analysis,
([79]). Applying the very important and useful *independence approximation* on
the behaviour of packets in packet switching networks he recommended to use
open exponential queueing networks (*Jackson networks*, [69]) to predict equilibrium queue lengths distributions, waiting times, and mean transmission times,
what is now generally termed *quality of service* for these systems.

H. Daduna: Queueing Networks with Discrete Time Scale, LNCS 2046, pp. 1–7, 2001.
© Springer-Verlag Berlin Heidelberg 2001

A second breakthrough was the invention of network models with different customer classes, class dependent random routing, and general transmission time distributions for some specific service regimes by Baskett, Chandy, Muntz, and Palacios. These networks are now well known as BCMP networks [4].
A more versatile class of network models including the BCMP networks was introduced at nearly the same time by Kelly [74], for a detailed introduction see [75]. Starting from both of these classes of network models a lot of generalizations have been introduced providing us with even more detailed modeling tools in continuous time, for a more recent survey see [127].

In the context of continuous time systems discrete time Markov chains play an important role. Defined as embedded chains they provide us with techniques to deal with continuous time single servers or even networks when the system is observed only at specific time points, e.g., arrival or departure instances. For an introduction see Kleinrock's monograph [81], Volume I, Part III, and for the application of these methods in performance analysis of computer and communication systems [82], Volume II.
Another use of discrete time models is to approximate continuous time models via time quantization, which especially serves for the numerics of direct computations, see e.g. [82], Volume II, Chapter 2.6.

While these applications of discrete time Markov chains to queueing models are to be considered as construction of auxiliary models for to investigate continuous time stochastic networks, another approach (of completely different character) to performance analysis of computer and communication systems was developed around 1965. This approach was triggered by the introduction of real world systems which show an inherent generic slotted time scale. In the field of computer science time–shared computing systems are perhaps the most well-known and earliest of such systems, investigated with methods from discrete time Markov chain theory, [78], [80]. Here the discrete time scale is prescribed by the size of the time slot that is given to a job before the processor is dedicated to the next job – serving jobs in a round–robin regime. A further prominent example from the area of communication networks is the slotted ALOHA protocol. This protocol is an example for a medium access protocol which governed the contention for shared bandwidth by different users in satellite transmission networks, see [142], chapter 6, and the references there.

It turned out that the technical difficulties which arise when working under a discrete time scale were in many cases considerably greater than in dealing with continuous time models. The reason for this is the combinatorial complexity which appears in the solution procedures for the global balance equations of these systems, [78], [80]. A consequence at that time was to use continuous time approximations for the generic discrete time systems, e.g., converting time–sharing systems into processor–sharing models. For a first highlight of this successful approach see [114], and for a review with more elaborated applications in performance evaluation (e.g.) [82], Volume II, Chapter 4.

Following this great success of the continuous time theory, discrete time queueing theory, especially the theory of queueing networks with explicit discrete

time scale, seem to lay dormant for about nearly twenty years. But a continuous –although thin– stream of papers on the subject appeared over the years:
For summaries and fundamental questions on the state–of–the–art at that time
we refer to the lectures of Kobayashi (in [95]), where applications in time–
multiplexing transmission systems are described, and the monograph of Hunter
([66] and [67]), where the fundamental single server queues are elaborated on.

The recent interest in discrete time queueing models emerged from the introduction of ATM (Asynchronuous Transfer Mode) as the multiplexing technique
for Broadband Integrated Services Digital Networks (B–ISDN). For a collection
of recent studies see [87]. These networks are featuring (at least) three levels with
different time scales: Call level, burst level, and cell level, where network access
control may be applied. The latter two levels can be and mostly are modeled
by discrete time systems: Single switches for the network as well as the whole
network itself.
The ATM switches on the cell level are modeled by discrete time queues, because
these systems are assumed to work synchronously on the basis of a smallest time
unit. The time scale is prescribed by the time needed to transmit just one cell.

The renewed interest is expressed by a continuously growing number of further research papers on discrete time queueing systems appearing in journals
dedicated to the fields of computer science, electrical engineering, operations
research and mathematics. Only recently there appeared the mentioned three
books on discrete time queues, [14], [128], and [142], and special issues of *Performance Evaluation: Discrete time models and analysis methods* and *Queueing
Systems and Their Applications: Advances in discrete time queues* were dedicated to the subject. The *Editorial Introductions* [133], [101] of these issues serve
to advertise for this class of models as being useful in predicting and explaining
systems' behaviour and for being profitable for applications and challenging in
theory.

The aim of these lectures notes is not to give an introduction into a general
discrete time queueing theory, but to highlight some (in my view) important and
interesting subjects of actual research which at the same time can be viewed as
being potentially applicable in the fields mentioned above. To a certain extent
well developed is now the theory of single node systems in discrete time and it
is impossible to survey the whole literature concerning isolated systems. Single
node queueing theory in discrete time is well documented because the recent
books provide us with detailed information [14], [128], [142], and [132], the latter with an emphasis on elementary introduction of transform methods.
There is now available a vast literature on discrete time models for ATM switches, see e.g. the books [14] (especially the sections on "Further Readings" there),
[128], and the recent survey [140]. In–depth studies of single node systems in discrete time with direct applications to modeling ATM systems are [62] and [131].

Dealing with single nodes in the first chapter of the book will therefore be
only introductory for building networks which I shall describe completely and in
detail. Nevertheless in chapter 2 when generalizing recent results some versatile
single node models will be presented which seem to be new.

The research on queueing networks in discrete time is still in a somewhat pre-mature state. Despite the work of Hsu and Burke [65], Walrand [135], Pujolle, Claude, and Seret [109], and Schassberger, Daduna [33], [118] in the 1980ies, only with the beginning of the 1990ies a continuously active research in the field started, searching for general explicit solutions for the performance measures of networks. And still up to now general results providing explicit performance measures for discrete time queueing networks are rare. An exception is chapter 12 [18] where a versatile class of models is described which are developed for continuous time and discrete time in parallel. Our third topic below is dedicated to this (chapter 6).

In the well–established evaluation methodology of continuous time systems the computation of performance measures as indicators of the quality of service (equilibrium joint queue length distribution for the nodes, end–to–end–delay for messages, throughput, etc.) relies almost always on finding *product form equilibrium* behaviour for the networks. That means: The network's nodes behave in equilibrium as if they are independent, the network is therefore *separable*. On the other hand, for queueing networks with slotted time scale performance evaluation up to now is usually done either by simulation, or by using direct general numerical procedures, or by applying heuristic approximations, see for example [14], chapter 4. Therefore the most challenging problem in my opinion is to develop a *product form calculus* similar to the one in the continuous time setting. What we should hope at the time being is finding versatile classes of network models which allow explicit computations yielding closed form solutions of the main performance indices. And what we shall expect is that there will be in the discrete time case a close connection too between explicit formulas for performance indices and a certain product form structure of these formulas. The latter will then indicate possibly locally balanced behaviour of the networks. Some first steps in this direction collected from the literature which may be considered as being landmarks in the broad and unexplored field shall be described in this lectures. Their main features will be discussed with respect to applications and directions of hopefully fruitful further research.

In the course of describing these models some generalizations and specifications will be presented as well, which widen their applicability in the concrete modeling process and may expand the usage of analytical methods in discrete time performance analysis.

The results to be presented can be divided into three groups:
(1) Linear networks (closed cycles and open tandems) of single server nodes under the First–Come–First–Served regime with Bernoulli input stream and geometrically distributed service times were investigated by Hsu and Burke [65], Pujolle, Claude, and Seret [109] (1986), Pestien and Ramakrishnan [107], and Daduna [26], [29], [28], [31]. We prove in chapters 3, 4 a product form equilibrium and compute the distribution of the end–to–end–delay. At present there seem to be severe limitations for generalizations of this class of networks. This point will be discussed.

(2) Networks of doubly stochastic and geometrical queues (which are discrete time analogues of Kelly's symmetric servers and of standard exponential queues): Customers of different types move through the network governed by a general routing mechanism and request for service according to distributions which depend on their type and the node where they stay. Under a restriction on simultaneous movements product form equilibrium is derived and the conditional expected system time for a customer is computed given his individual requests for service at the stages of his itinerary through network. This section reports on work done by Schassberger, Krüger, and Daduna, [33], [119], [89].

(3) The naturally emerging batch movements of customers and their batch service in discrete time stochastic networks suggest to consider these phenomena from a more abstract point of view. This has been done in Walrand's [135] early paper, and in a series of recent papers of Henderson, Taylor, Pearce, Northcote, Boucherie, van Dijk, Chao, Miyazawa, Pinedo, and others, see e.g.: [55], [60], and [98], [99]. Here the service and routing mechanism is either defined on the basis of an abstract transition scheme that generates explicit steady state distributions from its very definition, or the customers move nearly independent from one another through the system, which yields product form via the independent behaviour.

This research is partially summarized in chapter 12 of [18], and we deal with that topic in chapter 6, presenting some selected models.

There are further directions of research on queueing networks in discrete time not dealt with in these lecture notes. Some representatives may be found in the following references:

Morrison [103] considered a two–stage tandem with deterministic service times and interfering Bernoulli arrival processes. This was generalized by Boxma and Resing [11] to more general arrivals into an n–stage tandem. The results reported are mainly by means of transform methods.

To consider service systems with deterministic service times of duration of one time slot seems to be a natural task in the context of modeling ATM networks. This is because the size of the time slot is usually chosen to be the time needed to transmit a single cell. Turning to random transmission times follows the need to incorporate the load of competing parallel transmissions from different sources and the introduction of local traffic control (e.g. leaky bucket) which consumes extra time.

Sharma [124] proved a functional Central Limit Theorem and a functional Law of Iterated Logarithm for discrete time stochastic networks with overall deterministic–(1) service times and renewal arrivals from the exterior. An overview of Brownian and non–Brownian functional central limit theorems for single server discrete time queues is in [138].

Kouvatsos, Tabet–Aouel, and Denazis [88] apply maximum entropy principles to obtain product form approximations for the steady state probabilities of a rather general switching processs in a discrete time model.

A theoretical treatment related to the systems dealt with in this survey is presented by Henderson, Pearce, Taylor, and van Dijk [61] : They developed a discrete time *Generalized Semi–Markov* formalism similar to that in continuous time, [48], [115], [116]. The latter, although originally developed for rather theoretical purposes, is nowadays thought to be the adequate description for *Discrete Event Dynamic Systems* in stochastic simulation procedures.

Finally I want to stress that the application of the theory to modeling ATM networks relates the topics of the lecture notes only to one special class of application. There are numerous other applications, where a discrete time scale is essential for the system's behaviour and its description and understanding – and therefore for the modeling process. Some few examples from related fields are:

Cyclic–reservation protocols for high speed networks [134].

Woodward elaborates in his textbook [142] in detail medium access protocols for satellite networks (e.g. ALOHA protocols) and local area networks (under carrier sensing and token passing protocols) and their modelling by discrete time systems. He focusses on the network models invented by Walrand [135], see section 6.4. Similar models in a discrete time setting are described by Kobayashi [83] in his lecture notes.

Several discrete time network models for parallel and sequential task processing are developed by Hofmann, Müller, and Natarajan [63]. They compared the performance of the different processing schemes under different load behaviour.

General inventory models with deterministic lead times [72]: This suggests to use the lead time duration as a time unit. To tackle inventory problems on that basis in a similar way with methods described here seems to be promising. A completely different approach to use discrete time methods is suggested by Tempelmeier [130], [129]: Time unit is the minimal time between two successive inventory review points, often exactly one day. This coincides with often having exactly one arrival point for orders per day.

Conveyor and transportation models with equidistant carriers admit a natural discrete time scale, see for an evaluation with application to communication systems [1].

Population models in space and time, where time is scaled by the sequential arrivals of successive generations; discrete time compartment models (see [75] for the continuous time case).

A survey of applications of discrete time queueing systems is section 1.1 of [14] with a focus on communications systems, dam models, and transportation systems.

1.2 Symbols and Conventions

For ease of notation throughout the book we fix a common probability space $(\Omega, \mathcal{F}, P)$ and assume all random variables which occur to be defined on $(\Omega, \mathcal{F}, P)$, unless otherwise specified.

If a probabiliy measure P on $(\Omega, \mathcal{F})$ that governs some stochastic process

$$X = ((X_n : (\Omega, \mathcal{F}, P) \longrightarrow (E, \mathcal{S})) : n \in I\!N)$$

has to be specified further, often for guaranteing that X is stationary with stationary distribution (= steady state = equilibrium) π, this will be indicated by writing P_π. An expectation e.g. of X_n, with respect to such P_π is written $E_\pi X_n$. Similar expressions will be selfexplaining.

$I\!R$ denotes the real numbers, $I\!R_+ := [0, \infty)$.

The natural numbers are $I\!N := \{0, 1, 2, \dots\}$, the strict positive natural numbers are $I\!N_+ := \{1, 2, 3, \dots\}$, and we denote $Z\!\!\!Z := \{\dots, -2, -1, 0, 1, 2, \dots\}$.

We denote by

$$\delta(a, b) = \begin{cases} 1 & if \quad a = b \\ 0 & if \quad a \neq b \end{cases}$$

the Kronecker delta, and by

$$\eta(a, b) = \begin{cases} 1 & if \quad a \neq b \\ 0 & if \quad a = b \end{cases}$$

the complementary Kronecker delta.

Chapter 2

State Dependent Bernoulli Servers

In this section we consider the most elementary queueing system in discrete time. It is the analogue of the state dependent exponential single server queue in continuous time under First–Come–First–Served (FCFS) regime. We begin with systems where all customers are indistinguishable and summarize in section 2.1 the steady state performance measures. Most of those problems can be interpreted as special questions about the behaviour of random walks in discrete time. This does not hold in case of customers of different types arriving at the service node which we consider in section 2.2. In Section 2.3 we reconsider both models and allow immediate feedback of departed customers to the server.

2.1 Indistinguishable Customers

The time scale for our systems is $\mathbb{N} = \{0, 1, 2, \dots\}$, sometimes we shall use $\mathbb{Z} = \mathbb{N} \cup -\mathbb{N} = \{\dots, -2, -1, 0, 1, 2, \dots\}$. There is a single service facility where at each time instant at most one customer may be served. Customers are indistinguishable and arrive one by one randomly at the server. If at the arrival instant the server is free the service of the arriving customer immediately commences. Otherwise an arriving customer enters the waiting room which is organized on a FCFS basis (sometimes called FIFO: First–In–First–Out). If a customer has obtained his total service request he departs immediately from the system. If a customer departs and there is at least one further customer present then the customer at the head of the waiting line enters the server, his service commences immediately, and all other waiting customers are shifted one place up in the line. The time needed for reorganizing the queue is assumed to be neglectible (zero time).

The system's development over time will be described by a discrete time stochastic process $X = (X_t : t \in \mathbb{N})$ with state space $\mathbb{N}$. X_t denotes the

H. Daduna: Queueing Networks with Discrete Time Scale, LNCS 2046, pp. 9–22, 2001.
© Springer-Verlag Berlin Heidelberg 2001

Fig. 2.1. Bernoulli Server

number of customers present at time t, either in service or waiting, shortly the *queue length* at time t.

The randomness of the system is due to the following assumptions:
If at time t a customer is in service and if there are $n - 1 \geq 0$ other customers present then this service ends in the time segment $[t, t+1)$ with probability $p(n) \in (0, 1)$ and the customer will depart at the end of this time slot; with probability $q(n) = 1 - p(n)$ this customer will stay at least one further time quantum. The decision for a customer whether to stay or to leave is made independently of anything else other than the queue length at time t. If at time $t \in \mathbb{N}$ there are $n \geq 0$ customers present then at the end of time slot $[t, t+1)$ a new customer will arrive with probability $b(n)$; with probability $c(n) = 1 - b(n)$ there will be no arrival. The decision whether an arrival will occur or not is made independently of anything else other than the queue length at time t. Such an arrival stream will be termed henceforth *state dependent Bernoulli arrival process*. (For some special models we shall allow $p(n) = 1$ and/or $b(n) = 1$; see e.g. corollary 2.8, example 2.9.)

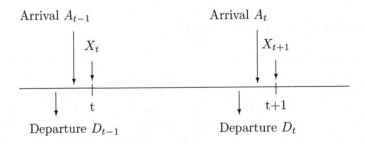

Fig. 2.2. Regulation of arrivals and departures

Unless otherwise specified, we shall assume throughout the notes the following rules for the regulation of simultaneous events: All arrivals (and departures) occur at the end of the respective time slots (*LA-rule*; late arrivals). If at the same epoch an arrival and a departure occur we always assume that the departure event takes place first. (*D/A-rule*; departure before arrival, see [51].) The state of the system is recorded at times $t \in \mathbb{N}$ just after possible departures and arrivals have happened.(Figure 2.2.)

A pathwise analysis of the system considers sequences of successive arrivals and departures which result in the queue length process observed at the time epochs at slot boundaries $t \in I\!N$.

Definition 2.1 (Departure and arrival process) *We denote by*

$$D = (D_t : t \in I\!N) \qquad and \qquad A = (A_t : t \in I\!N)$$

the sequences of numbers of departing, resp., arriving customers in time slot $[t, t+1)$*, and assume*

$$P(D_{t+1} = k | X_s, D_s, A_s : s \le t, X_{t+1}) = P(D_{t+1} = k | X_{t+1}), \qquad t \ge 0,$$

and

$$P(A_{t+1} = k | X_s, D_s, A_s : s \le t, X_{t+1}, D_{t+1}) = P(A_{t+1} = k | X_{t+1}), \qquad t \ge 0$$

to hold.

Corollary 2.2 (Queue length process) *The queue length process is pathwise defined by*

$$X_{t+1} = X_t - D_t + A_t, \quad t \in I\!N, \qquad X_0 \quad prescribed. \tag{2.1}$$

$X = (X_t : t \in I\!N)$ *is a time homogeneous Markov chain.*

In the literature there is a great variety of arrival and departure regimes defined. These regimes usually reflect physical behaviour of the systems under consideration and specification of protocols which govern the interaction of different processes runnning concurrently in the system. For the single server queue considered in this section an overview and a comparison of different regulation schemes is given by Hunter [67], more recent are [51], and paying more attention to the effect of different regulation schemes for simultaneous events in networks of queues on the performance analysis of customer oriented quality of service, [39], [38].

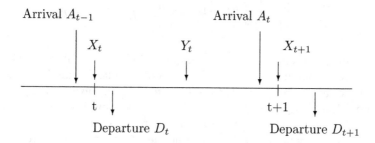

Fig. 2.3. Regulation of arrivals and departures in [18]

In [18], p.345, the regulation of arrivals and departures is pointed out to be essential for proving quasi–reversibility of the process. It is assumed that early departures occur at times $t+$ and arrivals enter the system at $(t+1)-$ in time slot $[t, t+1)$, see figure 2.3. The state process X is observed at slot boundaries t, and additionally a second state process Y is observed at times $t + 1/2$, as is done, e.g., in [51] as well.

It should be noticed that the pathwise construction of the queue length process in corollary 2.2 coincides with the queue length process constructed and observed at the boundaries of the time slots in [18], p.346, but the structure of the transition probabilities in definition 2.1 differ from the the internal structure (intermediate jump probabilities) of their transition mechanism which requires

$$P(A_{t+1} = k | X_s, D_s, A_s : s \leq t, X_{t+1}, D_{t+1}) = P(A_{t+1} = k | X_{t+1} - D_{t+1}), \quad t \geq 0$$

to hold. But nevertheless in many cases the overall transition probability for the state process X can be represented in either way.

The discrete time queueing system described so far is called a *state dependent Bernoulli server*. The term *Bernoulli server* stems from the case of state independent service and arrival probabilities (see corollary 2.7 below) and is generalized to our setting. The fundamental characteristics of the system are well known from the discrete time birth and death chain theory, see remark 2.4.

Theorem 2.3 (Steady state) *Let* $X = (X_t : t \in \mathbb{N})$ *with state space* $\mathbb{N}$ *denote the queue length process of the state dependent Bernoulli server described above.*
X is a homogemeous Markov chain.
If $p(n), b(n) \in (0, 1), \forall n \in \mathbb{N}$, then X is irreducible and aperiodic on $\mathbb{N}$.
If X is irreducible and aperiodic on $\mathbb{N}$, then X is positive recurrent, hence ergodic, if and only if

$$H := \sum_{n=0}^{\infty} \frac{\prod_{m=0}^{n-1} b(m)}{\prod_{m=0}^{n} c(m)} \cdot \frac{\prod_{m=1}^{n-1} q(m)}{\prod_{m=1}^{n} p(m)} < \infty. \tag{2.2}$$

If X is ergodic then its unique stationary and limiting distribution $\pi = (\pi(n) : n \in \mathbb{N})$ is

$$\pi(n) = \frac{\prod_{m=0}^{n-1} b(m)}{\prod_{m=0}^{n} c(m)} \cdot \frac{\prod_{m=1}^{n-1} q(m)}{\prod_{m=1}^{n} p(m)} \cdot H^{-1}, \quad n \in \mathbb{N}. \tag{2.3}$$

Proof : From the independence assumptions and the specification of the arrival and departure probabilities via queue lengths only it follows that X is a homogeneous Markov chain. From $p(n), b(n) \in (0, 1), n \in \mathbb{N}$, we have $P(X_{t+1} = n | X_t = n) > 0$ and $P(X_{t+|m-n|} = m | X_t = n) > 0$, hence X is aperiodic and irreducible.

The equilibrium equations $\mathbf{x} = \mathbf{x} \cdot \mathbf{p}$, where $\mathbf{p}$ is the one–step transition matrix of X and $\mathbf{x} = (x(n) : n \in \mathbb{N})$ is a nonnegative row vector, are solved by

$$x(n) = \frac{\prod_{m=0}^{n-1} b(m)}{\prod_{m=0}^{n} c(m)} \cdot \frac{\prod_{m=1}^{n-1} q(m)}{\prod_{m=1}^{n} p(m)}, \quad n \in \mathbb{N}.$$

Normalizing $\mathbf{x}$ yields the probability solution (2.3) if (2.2) holds. $\odot$

Remark 2.4 (Random walks in discrete time) *The queue length of the Bernoulli server, i.e., the Markov chain X, moves at most one step up or down in a time slot, i.e., it is a general random walk on $\mathbb{N}$ (with reflection at 0) in discrete time in the sense of [77], p. 84, or a birth and death chain in discrete time, see [66],(Vol. I), p. 178. Theorem 2.3 and the corollaries below are therefore simple consequences of the limiting and stationary behaviour of birth and death chains, see [67], (Vol. II), Example 7.2.2, p. 107.*

Remark 2.5 (Computational problems) *Due to allowing the general form of arrival and service probabilities it is in general not possible to give closed form solutions for steady state probabilities (even if these exist). This is in parallel to the continuous time birth and death process theory. Even determining recurrence and transience has to refer to the data of the specific system under consideration. For details see: Hunter [67], Vol. II, example 7.2.2, p. 107.*

An example with explicitly given norming constant is provided by systems with state independent arrival and service probabilities as described in the corollary 2.7 below. This yields a (homogeneous) random walk with reflection at 0.

Notice that the problem discussed here is different from the usual problem of computing norming constants for large closed queueing networks, see section 3.4. Both problems will concurrently occur e.g. in explicity evaluating the steady state distribution in theorem 3.2.

Remark 2.6 (Reversibility) *For the case of state independent arrival probabilities, i.e., if the arrival process is a Bernoulli–(b) stream, Hsu and Burke [65] proved that in steady state the queue length process X is reversible in time. This implies that the departure process in equilibrium is a Bernoulli–(b) process. Further, in equilibrium, the departure process up to t and the state at t are independent. This lead Hsu and Burke to apply separability properties to tandem queues.*

For random walks on $\mathbb{N}$ in discrete time the Bernoulli property of the departure process means, that even with state dependent death rates the downstep process is homogeneous if the birth probabilities are constant.

In [18], example 12.10, and the remark below on p.354, it is shown that this queue is quasi–reversible according to the definition 12.6 there. This again leads to the statement of Hsu and Burke on separably coupling such queues in a tandem.

The system dealt with in theorem 2.3 is neither reversible nor quasi–reversible.

Corollary 2.7 (State independent Bernoulli server) *Assume that in the setting of theorem 2.2 we have $p(n) = p \in (0,1)$, and $b(n) = b \in (0,1), n \in \mathbb{N}$. Then X is ergodic if and only if $b < p$, and if this holds the stationary distribution of X is*

$$\pi(n) = (1 - \frac{b}{p})\left(\frac{bq}{cp}\right)^n \left(\frac{1}{q}\right)^{\eta(0,n)}, \qquad n \in \mathbb{N}. \tag{2.4}$$

The arrival process is a Bernoulli process with probability b for successes and the service times are geometrically distributed on $\mathbb{N}_+ = \{1, 2, 3, \ldots\}$ with parameter p. The interarrival times are geometrically distributed, and the service process can be thought to be regulated by a Bernoulli-(p) process.

Weakening the assumptions on the arrival and service probabilities leads to some interesting special cases.

Corollary 2.8 (Loss systems) *Assume that in the setting of Theorem 2.2 we have $b(n) \in (0,1), n \leq L - 1 > 0, b(n) = 0, n \geq L$.*
Then X is ergodic on $E = \{0, 1, \ldots, L\}$, and the stationary distribution of X is $\pi = (\pi(n) : n \in E)$ given by

$$\pi(n) = \frac{\prod_{m=0}^{n-1} b(m)}{\prod_{m=0}^{n} c(m)} \cdot \frac{\prod_{m=1}^{n-1} q(m)}{\prod_{m=1}^{n} p(m)} \cdot H^{-1}, \quad n \in \mathbb{N}, \tag{2.5}$$

where

$$H := \sum_{n=0}^{L} \frac{\prod_{m=0}^{n-1} b(m)}{\prod_{m=0}^{n} c(m)} \cdot \frac{\prod_{m=1}^{n-1} q(m)}{\prod_{m=1}^{n} p(m)}$$

is the norming constant.

Example 2.9 (Deterministic service times) *The case of deterministic service times is of specific interest, because in modelling transmission lines where cells with constant length are to be transmitted this constant cell length τ constitutes a generic discrete time quantum such that $\mathbb{N} \cdot \tau = \{0, \tau, 2\tau, 3\tau, \ldots\}$ is a suitable time scale for a process reflecting the development of the system over time. Assuming $\tau = 1$ and $p(n) = 1, n \in \mathbb{N}_+$, and starting the system with $P(X_0 = n) > 0$ for all $n \in \mathbb{N}$, we find that X is not irreducible on $\mathbb{N}$.*
Indeed, if $b(n) \in (0,1)$ for all $n \in \mathbb{N}$, then X is irreducible and ergodic on $E = \{0, 1\}$. The stationary distribution π is given by

$$\pi(0) = \frac{1}{c(0)} \cdot H^{-1}, \quad \pi(1) = \frac{b(0)}{c(0)c(1)} \cdot H^{-1},$$

which formally fits into (2.3).
If $b = 1$ and $P(X_0 = n) > 0, \forall n \in \mathbb{N}_+, P(X_0 = 0) = 0$, then X is stationary for any such initial distribution.

The discrete time counterpart to the classical $M/M/s/\infty$ queue with s identical service channels under FCFS poses surprisingly many problems with respect to analytical evaluation. No simple closed form expressions for the steady state seem to be at hand. Usually numerical procedures with root solving for multidimensional boundary equations are applied. Models related to that problem are dealt with e.g. in [15], [36] [126], [14], section 4.1.2, where the service time is deterministic 1 while the arrival streams are more general. The case of no waiting room is e.g. considered in [20]. The complexity of problems and the numerical difficulties which arise in using these multiserver queues in discrete time for modeling controled ATM switches are described e.g. in [112], where a leaky–bucket control is investigated in detail.

Pestien and Ramakrishnan [107], section 3, proved that including these servers into a closed cycle of geometrical queues destroyes the product form equilibrium if $s < \infty$.

It is well known that in continuous time for the $M/M/s/\infty$ queue there exists an equivalent state dependent single server queue with the same steady state distribution. From the discussion above and the explicit steady state (2.3) it follows that such an equivalence is not possible in discrete time. However it is for a subclass of $M/M/s/\infty$ queues possible to construct a single server approximation in a rather direct way.

Example 2.10 (Multiserver queue approximation in low traffic)
Consider an $M/M/s/\infty$ queue in discrete time with Bernoulli arrival stream of intensity $b \in (0,1)$ and service probability $p \in (0,1)$ such that $b < p \cdot s$ holds. The conditional service intensity for that system (mean number of departures per time unit given the number of customer in system at the begin of a time slot) is the same as that of a single server state dependent queue according to theorem 2.3 with service probabilities

$$p(n) := p \cdot \min(n, s).$$

Studying the behaviour of individual customers with respect to their delay behaviour (waiting time and sojourn or passage time distribution) in equilibrium is a main topic in performance analysis. A first step towards this is determining under equilibrium conditions the state distribution at arrival instants, i.e., what an arriving customers observes just before he enters the system. To be more specific: An arriving customer observes at his arrival at $t-$ exactly n other customers present, if at time t $X_t = n+1$ holds and an arrival occurs. A customer who is to be observed during his stay in the node for determining (probabilistic) characteristics of his behaviour is usually attributed to be a test customer. To deal with this notion and then to derive from a test customer's behaviour conclusions about the behaviour of a typical customer needs care because e.g. knowing that a test customer commences his sojourn in the system prevents in general the system from being in equilibrium. The theoretical framework for dealing with this problem is Palm calculus of point process theory (see [2]) which fortunately enough can be proved to reduce to investigate elementary conditional probabilities in discrete time models (see [2], section 7.4). We formulate this as arrival theorems for customers, usually starting from a system in equilibrium.

Theorem 2.11 (Arrival theorem) *Let the queue length process X of a state dependent Bernoulli server be in steady state π according to (2.3). Consider for $t > 0$ the event*

$A(t) = \{$ *At time t a new customer arrives at the system* $\}$. *Then for* $n \in I\!N$

$$\pi_1(n) := P(X_t = n + 1 | A(t)) = \frac{\prod_{m=0}^{n} b(m)}{\prod_{m=0}^{n+1} c(m)} \cdot \frac{\prod_{m=1}^{n} q(m)}{\prod_{m=1}^{n} p(m)} \cdot H_1^{-1}, \qquad (2.6)$$

is the probability (with normalization constant H_1) that conditioned on an arrival at time t the arriving customer finds n other customers before him in service or waiting.
We denote by $\pi_1 = (\pi_1(n) : n \in I\!N)$ the arrival probability. *The arrival state and arrival probability refer to the disposition of the other customers present, the new arrival is not counted.*

Proof : For $t > 0$ we compute elementary conditional probabilities $P(X_t = n + 1 | A(t)) = \sum_{r=0}^{\infty} P(X_t = n + 1, X_{t-1} = r, A(t)) P(A(t))^{-1}$ ⊙

The common interpretation of π_1 is that it describes the distribution of the other customers' disposition in an arrival instant at the node under equilibrium conditions, the jumping customer not counted. Remarkable is that this arrival distribution has not the form of the equilibrium of the system even if the arrival process is a state independent Bernoulli process.. In continuous time it is true for the case of state independent arrivals that the time stationary and customer arrival stationary distribution coincide. For open systems with external Poisson arrivals this is the celebrated PASTA property (Poisson Arrivals See Time Averages) [141].
From a general point of view the PASTA theorem and its successors and relatives determine the stationary and asymptotic distribution of systems when the observation points are prescribed by an associated (embedded) point process. Comparing the stationary distribution of the embedded state process with the time stationary distribution of the systems (seen by an outside observer) is the topic of many research activities in queueing theory. Since Wolff's PASTA theorem [141] appeared the research in this field yielded many generalizations of that property. These have been proved and popularized under names like ASTA, EPSTA, MUSTA. For a review see [2], Chapter 4, Section 3, or [84], and [42], Chapters 3,4. Using point process terminology, EPSTA and its relatives are concerned with properties of special Palm measures of stationary point processes: Papangelou's formula, which connects the point process intensity and time stationary expectations with Palm stationary expectations, allows the derivation of a variety of formulas summarized under the title *job observer properties* in single node systems as well as in networks of queues. For continuous time, see e.g. [2], Example 3.2.2, [97], Example 3, [42], Section 4.3, and [35].
Starting from Palm theory in discrete time ([2], Chapter 1, Section 7.4) a similar development is possible. Palm measures in discrete time are expressed as elementary conditional probabilities. This makes the theory more elementary although explicit computations are tedious, see sections 3.4 and 3.5.
  In discrete time, however, a PASTA analogue usually does not hold, although exceptions can be found under certain conditions. An early result was proved by

Halfin in [52]. Characterisation theorems of the PASTA type (thereby strengt-hening the BASTA–results (Bernoulli Arrivals See Time Averages) from [96]) were proved by El-Taha and Stidham [41], (see also [42], Section 2, Theorem 3.18 and Corollary 3.19. Miyazawa and Takahashi [102] proved ASTA in a di-screte time point process setting by using a rate conservation principle. They also observed that for some natural systems this property does not hold. In this notes we prove "arrival theorems" for the different cases, where individual cu-stomers' behaviour when entering the system is of interest. These provide us with the necessary basic information to compute customer oriented performance measures. The proofs can be performed either by applying the general theory or by directly evaluating elementary conditional probabilities, i.e. Palm probabili-ties in discrete time processes. The latter approach is generally chosen here. We assume in any case that the Markov process which describes the time evolution of the system is stationary.

Note: If we rescale time, approaching in the limit a continuous time scale, our systems transform to exponential queueing system: Then in the limit the distinction between arrival distribution and steady state of the system disappears for state independent arrival probabilities.

Because customers are scheduled according to FCFS regime the arrival theo-rem provides us with information about the number of services which will be performed before a test customer, who enters the system in equilibrium, will be served. This enables us in principle to write down the end–to–end–delay time distribution. But even for this simple single node, single server system explicit results seem to be not known in full generality. We consider the case of state independent service probabilities:

Theorem 2.12 (End–to–end–delay) *Consider the Bernoulli server with state dependent arrival rates* $b(n) \in (0,1)$ *and state independent service rates* $p(n) = p \in (0,1)$ *in equilibrium with a test customer arriving at time 0 finding the other customers distributed according to* π_1*, see (2.6). We denote by* P_{π_1} *a probability measure on* $(\Omega, \mathcal{F})$ *which governs the development of* X *under this conditions and by* $E_{\pi_1}[\cdot]$ *expectations under* P_{π_1}*.*
Denote by S *this test customer's total sojourn time in the system and by*

$$E_{\pi_1}\theta^S = \sum_{s=0}^{\infty} P_{\pi_1}(S=s)\theta^s, \quad |\theta| \leq 1,$$

the generating function (of the distribution) of S*. If the generating function of the arrival factors in* π*, see (2.3), is*

$$\alpha(\theta) = \sum_{n=0}^{\infty} \frac{\prod_{m=0}^{n-1} b(m)}{\prod_{m=0}^{n} c(m)} \theta^n, \quad |\theta| \leq q/p, \tag{2.7}$$

then

$$E_{\pi_1}\theta^S = \frac{\alpha(\frac{q\theta}{1-q\theta}) - \alpha(0)}{\alpha(\frac{q}{p}) - \alpha(0)}, \quad |\theta| \leq 1. \tag{2.8}$$

Proof : We have by conditioning

$$E_{\pi_1}\theta^S \quad = \quad \sum_{n=0}^{\infty} P_{\pi_1}(X_0 = n+1)E_{\pi_1}\left[\theta^S | X_0 = n+1\right]$$

$$\overset{(1)}{=} \quad \sum_{n=0}^{\infty} \frac{\prod_{m=0}^{n} b(m)}{\prod_{m=0}^{n+1} c(m)} \cdot \left(\frac{q}{p}\right)^n \cdot H_1^{-1} \cdot \left(\frac{p\theta}{1-q\theta}\right)^{n+1}.$$

In $\overset{(1)}{=}$ we used that the service times of the customers are geometrically distributed with parameter p on $I\!N_+$ and independent. From $H_1 = (p/q)(\alpha(q/p) - \alpha(0))$
the result follows immediately. $\odot$

2.2   Customers of Different Types

In this section we consider the service system of section 2.1 and assume that
customers arriving at the system may be of different types. All customers are
served according to the same rules (FCFS) and have the same distribution for
their requested amount of service. Later on the type description will determine
a customer's behaviour and service in networks of state dependent nodes, see
section 5.3. The details:

The node characteristics remain the same as in the previous section. Customers
of different types arrive according to a state dependent Bernoulli process at the
node, are served according to FCFS, and thereafter depart from the system.

Joint arrivals and departures are scheduled according to LA-D/A regime (*late
arrivals* and *departure before arrivals*), see [51].

We assume that there is a single chain of customers, all customers share the
same type set M. The entrance type of an arriving customer is chosen according
to the following rules: The external arrival probabilities depend on the history
of the system only through the actual total population size of the system and on
the type of the arrival, i.e., if at time $t \in I\!N$ there are n customers present, then
a new arrival of type m appears in $(t, t+1]$ with probability $b(n) \cdot a(m) \in (0,1)$.

Departure and arrival decisions are conditionally independent given the actual vector of queue lenghts.

A typical state of the system is descibed by a type sequence $x = (x_1, \ldots, x_n) \in$
M^n, where for $n > 0$ x_1 is the type of the customer in service, x_2 is the type of
the customer at the head of the queue, $\ldots$, x_n is the type of the customer who
arrived most recently. The empty system is denoted by $x = e$. (For definiteness
we set for the empty system the queue length $n = 0$.) These states are sufficient
for constructing the state space of the system where X is living on.

Let $X(t)$ denote the state of the node at time $t, t \in I\!N$. $X = (X(t) : t \in I\!N)$
is a discrete time Markov chain with state space $\tilde{S} := \{e\} \cup \bigcup_{n=0}^{\infty} M^n$ and
transition matrix $p = (p(x,y) : x, y \in \tilde{S})$. X is irreducible on $\tilde{S}$. The problem of
stabilisation for this tandem system is solved by the following theorem.

Theorem 2.13 (Steady state) *The Markov chain X is ergodic if and only if*

$$H = \sum_{(n \in I\!\!N)} \left(\frac{\prod_{m=0}^{n-1} b(m)}{\prod_{m=0}^{n} c(m)} \right) \left(\frac{\prod_{m=1}^{n-1} q(m)}{\prod_{m=1}^{n} p(m)} \right) < \infty.$$

If this ergodicity condition is fulfilled, then the unique equilibrium distribution of X is $\pi = (\pi(x) : x \in \tilde{S})$ given by

$$
\begin{aligned}
\pi(x) &= \pi(x_1, \ldots, x_n) \\
&= \left(\frac{\prod_{m=0}^{n-1} b(m)}{\prod_{m=0}^{n} c(m)} \right) \left(\prod_{k=1}^{n} a(x_k) \right) \left(\frac{\prod_{m=1}^{n-1} q(m)}{\prod_{m=1}^{n} p(m)} \right) \cdot H^{-1}, \quad (2.9) \\
x &= (x_1, \ldots, x_n) \in \tilde{S}.
\end{aligned}
$$

Proof : By inserting formula (2.9) into the equilibrium equation. We shall consider the case of networks of such nodes in detail below. $\odot$

Remark 2.14 (Steady state decomposition) *Similarly to theorems 2.3, 2.11, we observe a decomposition (separation) of the steady states into factors concerning the arrival probabilities, the service probabilities, and in addition the type selection probabilities. This separability is common to almost all product form steady states in continuous time and will occur later on in the framework of discrete time queueing networks as well, see e.g. theorems 3.2, 3.5, 4.1, 5.4, and 5.9. The separability opens the way to search successfully for explicit performance measures.*

Our aim is to investigate an individual customer's delay behaviour. We must therefore incorporate the types into an arrival theorem similar to theorem 2.11.

Theorem 2.15 (Arrival Theorem) *Consider the state process X of the node in equilibrium and denote by*
$A(m,t) = \{$at time t a customer of type m arrives at the node$\}$
the arrival event of interest. Then

$$
\begin{aligned}
\pi_{1,m}(x) &= P(X(t) = ((x_1, \ldots, x_n), m) | A(m,t)) \quad (2.10) \\
&= \left(\frac{\prod_{m=0}^{n} b(m)}{\prod_{m=0}^{n+1} c(m)} \right) \left(\prod_{k=1}^{n} a(x_k) \right) \left(\prod_{m=1}^{n} \frac{q(m)}{p(m)} \right) H_1^{-1}, \\
x &= (x_1, \ldots, x_n) \in \tilde{S},
\end{aligned}
$$

H_1 *is the norming constant, which does not depend on the type of the arriving customer.*

For the interpretation of $\pi_{1,m}$ see the remark following theorem 2.11. Note that the arrival distribution does not depend on the typ of the arriving customer. This is due to the fact that we consider an "open" system. For "closed" systems this property does not hold.

Corollary 2.16 (End–to–end–delay) *Consider the Bernoulli server with state dependent arrival rates $b(n) \in (0,1)$ and state independent service rates $p(n) = p \in (0,1)$ in equilibrium with a test customer of type m arriving at time 0 finding the other customers distributed according to $\pi_{1,m}$, see (2.10). We denote by $P_{\pi_{1,m}}$ a probability measure on $(\Omega, \mathcal{F})$ which governs the development of X under this conditions and by $E_{\pi_{1,m}}[\cdot]$ expectations under $P_{\pi_{1,m}}$.*
Denote by S this test customer's total sojourn time in the system. Then the generating function of the distribution of S is

$$E_{\pi_{1,m}}\theta^S = \sum_{s=0}^{\infty} P_{\pi_{1,m}}(S=s)\theta^s = \frac{\alpha(\frac{q\theta}{1-q\theta}) - \alpha(0)}{\alpha(\frac{q}{p}) - \alpha(0)}, \quad |\theta| \le 1. \qquad (2.11)$$

Here the $\alpha(\cdot)$ is the generating function of the arrival factors, given in (2.7).

Proof : Because the arrival distribution does not depend on the individual type of the test customer, the factors concerning types cancel and we end with computations similar to those in the proof of theorem 2.11. $\odot$

2.3   Bernoulli Servers with Immediate Feedback

Feedback queues are standard models in continuous time systems to model repeated visits of an item to a production or service facility. Another situation with typical feedback structure is rework which occurs due to production control at the exit point of a production stage. A class of network models which encompass these features are re–emtrant lines which are surveyed in [90] in the continuous time setting, and which are to be described for the discrete time case in section 4.4.

Applications in the realm of ATM transmission systems are described in [113] where in a node with service time deterministic–(1) the feedback mechanism models the successive transmission of cells of a message, the length of which is geometrically distributed. An important consequence is that introducing the feedback destroys the FCFS structure of the systems, which reflects real systems' protocol behaviour. In fact, the queueing regime used by the authors is what is called *Round–Robin* queueing discipline, and which is described in [93], see [92] and the refences there. We shall describe here the single feedback station in isolation, which will be used as the nodes of tandems to be considered in section 4.4.

We consider a state dependent Bernoulli server under FCFS as described in section 2.1 with indistinguishable customers and introduce a Bernoulli switch at the departure point of the node. A customer departing from the queue leaving behind $m-1$ customers is fed back into the waiting room (to the tail of the queue) with probability $r(m)$. If he was the only customer present he will obtain immediately a further service, otherwise he will join the tail of the queue. With probability $1 - r(m)$ he will leave the system. The decision whether to leave or to reenter the node is made independently of anything else. The regulation of customer movements in case of multiple events is: Departure before arrival for a joint arrival and departure (D/A) and feedback before arrival for a joint feedback and arrival (F/A).

The queue length process $X = (X_t : t \in I\!N)$ with state space $I\!N$ is Markovian. If $0 < b(\cdot), p(\cdot) < 1$, then X is irreducible and aperiodic on $I\!N$. A short reflection shows:

Observing the queue length process at times $t \in I\!N$ only is not sufficient to decide whether a feedback happened or no arrival *and* no departure happened or one arrival *and* one departure. We can determine from observing a path of X only single external arrivals and single departures to the external. The transition probabilities of the feedback queue are therefore

$$
\begin{aligned}
p(0,0) &= c(0), & &\text{(2.12)}\\
p(0,1) &= b(0),\\
p(0,m) &= 0, \quad m \neq 0,1,\\
p(n,n-1) &= c(n)p(n)(1-r(n)), \quad n \geq 1,\\
p(n,n) &= c(n)(q(n)+p(n)r(n))+b(n)p(n)(1-r(n)), \quad n \geq 1,\\
p(n,n+1) &= b(n)(q(n)+p(n)r(n)), \quad n \geq 1,\\
p(n,m) &= 0, \quad n \geq 1, m \neq m, m-1, m+1.
\end{aligned}
$$

These transition probabilities are identically to those of a state dependent Bernoulli server without feedback, with arrival probabilities $b(\cdot)$, and with service probabilities $p'(n) = p(n)(1-r(n)), q'(n) = 1 - p(n)(1-r(n)), n \in I\!N$. Therefore the Markovian queue length processes of these different systems are stochastically indistinguishable, and we conclude that theorem 2.3 applies for computing the steady state of the feedback queue.

Corollary 2.17 (Steady state of the feedback queue) *Let $X = (X_t : t \in I\!N)$ with state space $I\!N$ denote the queue length process of the state dependent Bernoulli server with immediate feedback.*
If X is irreducible and aperiodic on $I\!N$, then X is positive recurrent, hence ergodic, if and only if

$$
H := \sum_{n=0}^{\infty} \frac{\prod_{m=0}^{n-1} b(m)}{\prod_{m=0}^{n} c(m)} \cdot \frac{\prod_{m=1}^{n-1}(q(m)+p(m)r(m))}{\prod_{m=1}^{n} p(m)(1-r(m))} < \infty.
$$

If X is ergodic then its unique stationary and limiting distribution $\pi = (\pi(n) : n \in I\!N)$ is

$$
\pi(n) = \frac{\prod_{m=0}^{n-1} b(m)}{\prod_{m=0}^{n} c(m)} \cdot \frac{\prod_{m=1}^{n-1}(q(m)+p(m)r(m))}{\prod_{m=1}^{n} p(m)(1-r(m))} \cdot H^{-1}, \quad n \in I\!N. \quad\text{(2.13)}
$$

Note that even in the case of state independent arrival, departure, and feedback probabilities the effective arrival stream at the waiting room, i.e., the superposition of the external arrival stream and the feedback stream, is neither geometrical nor a state dependent Bernoulli arrival process as defined in section 2.1.

We consider now a feedback node where customers of different types from the set M of possible types are served, i.e., the model of section 2.2 with a queue length dependent Bernoulli feedback as described before. The state space for a Markovian description is $\tilde{S} = \{e\} \cup \bigcup_{n=1}^{\infty} M^n$, as defined before theorem 2.13. We notice that we usually can realize the occurrence of a feedback by a resulting permutation of the customers. We even can decide when a batch arrival due to a simultaneous feedback and external arrival appears. The only exception: All customers present are of the same type.

Consequently the transition probabilities of the feedback node have no longer an interpretation in terms of a Bernoulli node without feedback and suitably adjusted service probabilities. In the light of this the following result is somewhat surprising. The proof is straight forward but tedious checking the balance equations.

Theorem 2.18 (Feedback queue with different customer types) *Let $X = (X_t : t \in \mathbb{N})$ with state space $\tilde{S}$ denote the Markov chain describing the evolution of the state dependent feedback queue with different customer types. If $p(n), b(n) \in (0,1)$, and $r(n) < 1$ for all $n \in \mathbb{N}$, then X is irreducible and aperiodic on $\tilde{S}$.*
If X is irreducible and aperiodic, then X is positive recurrent, hence ergodic, if and only if

$$H := \sum_{n=0}^{\infty} \frac{\prod_{m=0}^{n-1} b(m)}{\prod_{m=0}^{n} c(m)} \cdot \frac{\prod_{m=1}^{n-1}(q(m) + p(m)r(m))}{\prod_{m=1}^{n} p(m)(1 - r(m))} < \infty.$$

If X is ergodic then its unique stationary and limiting distribution $\pi = (\pi(x) : x \in \tilde{S})$ is

$$\pi(x_1, \dots, x_n) \qquad\qquad (x_1, \dots, x_n) \in \tilde{S}, \qquad\qquad (2.14)$$

$$= \frac{\prod_{m=0}^{n-1} b(m)}{\prod_{m=0}^{n} c(m)} \cdot \prod_{k=1}^{n} a(x_k) \cdot \frac{\prod_{m=1}^{n-1}(q(m) + p(m)r(m))}{\prod_{m=1}^{n} p(m)(1 - r(m))} \cdot H^{-1}.$$

We obtain the round–robin discipline dealt with in [113], where service time is deterministic–(1) and the total system time being geometrically distributed, by putting $p(n) = 1$ for all $n \geq 1$ and the feedback probability being state independent as well. Nevertheless, the model in [113] includes several further modeling features not represented in the simplified feedback Bernoulli server here. In [93], [92] it is assumed that the number of packets arriving per slot is determined by a sequence of identically distributed random variables (batch Bernoulli arrival process), the service time is of (discrete) phase type, and the service mechanism is round–robin. Due to the general arrival process the steady state of the queue length process is no longer of product form, as it is in case of state dependent single arrivals of different customer classes [32].

Chapter 3

Closed Cycles of State Dependent Bernoulli Servers with Different Customer Types

In this chapter we study networks of Bernoulli servers in discrete time: Closed cycles with a fixed number of customers cycling. These are the analogues of cycles of exponential single server queues in continuous time under FCFS–regime, [50], [85].

We construct cycles with state dependent node behaviour and multiple customer types as described in section 2.2 and give references to the more classical results as corollaries. In section 3.1 we determine steady state behaviour and the individual customers' behaviour at arrival instants at the nodes. We concentrate on the unichain case throughout the notes because it reduces complexity of the presentation. The multichain case can be discussed along the same lines but with much more involved notation. The principle differences are sketched in section 3.2.

In section 3.3 we determine individual sojourn time distributions at the nodes for a travelling customer throughout his passage through a cycle. We end with explicit expressions for the generating function (z-transform) of the vector of the successive sojourn times which can be inverted easily by direct methods. Although these results are explicit, it turns out that similar to the continuous time case there are problems with numerical evaluation. We sketch in section 3.4 algorithms to compute the norming constants which occur in the probabilities obtained before.

Of historical interest is the time lag in the development of classical results, a remark on this may be in order here: The first explicit result on the steady state behaviour for closed cycles of state independent Bernoulli servers appeared

H. Daduna: Queueing Networks with Discrete Time Scale, LNCS 2046, pp. 23–52, 2001.
© Springer-Verlag Berlin Heidelberg 2001

in 1994 – see [107] and [106]. The parallel result for open series of such nodes appeared already 1976, see [65]. An explanation for the time lag between finding these parallel results is the combinatorial complexity of the closed system which cannot be escaped by proving time reversibility of the nodes' local state process in equilibrium as it was done in the open network case by Hsu and Burke. The direct proof of the steady state result by Pestien and Ramakrishnan [107] will be substituted here by considering the time reversed network process in conjunction with the original process (theorem 3.2 and proposition 3.3).

3.1   Steady State Behaviour and Arrival Theorem

We consider a closed cycle of state–dependent Bernoulli servers under FCFS queueing regime with unlimited waiting room. There are J nodes in the cycle, numbered $1, 2, \ldots, J$; customers leaving node j proceed immediately to node $j + 1$. (We formally define for the numbering of the nodes $J + 1 := 1$ and $1 - 1 := J$.)

K customers cycle in the system being of specific types $m \in M, 0 < |M| < \infty$. Customers may change their types according to a Markovian selection rule when departing from a node. All customers show the same behaviour with respect to changing their types. (The unichain case.)

With probability $r(i; m, m') \geq 0$ a customer of type m on leaving node i becomes a customer of type $m' \in M$ when entering node $i+1, i = 1, \ldots, J$. Given i and m, the selection of m' is done independently of anything else in the past of the system. We do not put further assumptions on the probability matrices $r(i) = (r(i; m, m') : m, m' \in M)$, so e.g. cyclic or deterministic behaviour of the type change process is possible. We only assume that the system of equations

$$\eta(i; m) = \sum_{m' \in M} \eta(i - 1; m')r(i - 1; m', m), i = 1, \ldots, J, m \in M, \qquad (3.1)$$

has a unique stochastic solution $\eta = (\eta(i; m) : i = 1, \ldots, J, m \in M)$.

The evolution of the system will be described by a multivariate discrete time stochastic process $X := (X(t) : t \in I\!N)$ as follows: Let $M(i) = \{m \in M : \eta(i; m) > 0\}$ denote the set of possible types which customers may show when staying in node $i, i = 1, \ldots, J$. A typical state of the system is denoted by $x = (x_1, \ldots, x_J)$, where $x_j = e_j$ or $x_j = (x_{j1}, \ldots, x_{jn_j}) \in M(j)^{n_j}, 1 \leq n_j \leq K, j = 1, \ldots, J$, and $n_1 + \cdots + n_J = K$. x_j is called a local state for node j with the following meaning:

If $x_j = e_j$ then node j is empty and we set $n_j = 0$.

If $x_j = (x_{j1}, \ldots, x_{jn_j}), n_j > 0$, then a customer of type x_{j1} is in service at node j, x_{j2} is the type of the customer waiting at the head of the queue, $\ldots$, and x_{jn_j} is the type of the customer who arrived most recently at node j. These local states are sufficient for constructing the state space of the system

$$\tilde{S} := \tilde{S}(K, J) :=$$

$$:= \{(x_1, \ldots, x_J) : x_j = (x_{j1}, \ldots, x_{jn_j}) \in M(j)^{n_j}, j = 1, \ldots, J, n_1 + \cdots + n_J = K\},$$

where $X = (X(t) := (X_1(t), \ldots, X_J(t)) : t \geq 0)$ is living on.

We assume that the nodes operate independently as single channel state dependent Bernoulli servers under FCFS with ample waiting room for all customers cycling. If at time t at node j a customer is in service and if there are $h - 1 \geq 0$ other customers present at that node then this service ends in the time segment $[t, t+1)$ with probability $p_j(h) \in (0,1)$ and the departed customer will be at the end of the queue of node $j + 1$ at time $t + 1$; with probability $q_j(h) = 1 - p_j(h)$ this customer will stay at least one further time quantum at node $j, j = 1, \ldots, J$. The decision for a customer whether to stay or to leave is made independently of the history of the system given the local actual state of X at j. A customer arriving at node $j + 1$ at time $t + 1$ will either join the end of the queue there (if other customers are present) or immediately enter service (if at time t node j was empty or there has been exactly one customer who obtained his last quantum of service time). If at some node at the same epoch an arrival and a departure occur at the same node we always assume that the departure event takes place first. (D/A–rule; departure before arrival, see [51].)

The state of the system is recorded at times $t \in I\!N$ just after possible departures and arrivals had happened. (Note that due to the ample waiting room assumption the A/D–rule (arrival before departure) will yield the same steady state distribution of the discrete time state process.

$\tilde{S}(K, J)$ is not necessarily the minimal state space. The set of those states which X will enter with positive probability may be a subset of $\tilde{S}(K, J)$, depending on the initial state of the process. This is due to the cyclic structure of the system which prevents customers from overtaking one another. We therefore assume that the following holds.

Assumption 3.1 (Irreducibility) *Depending on the initial state X is irreducible on its state space. We denote the state space in any case by $\tilde{S}(K, J)$, the meaning of which will always be clear from the context.*

Theorem 3.2 (Steady–State Distribution) *The Markov chain $X = (X(t) : t \in I\!N)$ is positive recurrent and its unique stationary distribution $\pi^{K,J} = (\pi^{K,J}(x) : x \in \tilde{S}(K, J))$ is given by*

$$\pi^{K,J}(x_{11}, \ldots, x_{1n_1}; \ldots; x_{J1}, \ldots, x_{Jn_J}) \tag{3.2}$$

$$= \prod_{j=1}^{J} \left(\prod_{k=1}^{n_j} \eta(j; x_{jk}) \right) \left(\frac{\prod_{h=1}^{n_j - 1} q_j(h)}{\prod_{h=1}^{n_j} p_j(h)} \right) \cdot G(K, J)^{-1}$$

$$(x_{11}, \ldots, x_{Jn_J}) \in \tilde{S}(K, J),$$

where $G(K, J)^{-1}$ is the norming constant.

Proof: From finiteness of the state space and the assumption on the unique stochastic solution of (3.1) positive recurrence follows. Instead of solving the

equilibrium equations directly the rest of the proof is done by considering the
time reversal of X and hence part of the proof of the following proposition [31].

$$\odot$$

Proposition 3.3 (Time Reversed Chain) *Denote by* $X = (X(t) : t \in \mathbb{Z})$
the stationary continuation of X *under* $\pi^{K,J}$ *on time axis* $\mathbb{Z}$ *as well. Then the
time reversal of* X

$$\bar{X} = (\bar{X}(t) := X(-t) : t \in \mathbb{Z}),$$

*describes a closed cycle of state–dependent single server nodes with the same
node characteristics as the original system but with* K *customers cycling in the
reversed direction. The type set of the customers is the same and type changes are
governed by Markovian matrices* $\bar{r}(i) = (\bar{r}(i; m', m) : m', m \in M), \quad i = 1, \dots, J,$
given by

$$\bar{r}(i + 1; m', m) := r(i; m, m') \cdot \frac{\eta(i; m)}{\eta(i + 1; m')}. \tag{3.3}$$

For state $x = (x_1, \dots, x_J) \in \tilde{S}(K, J)$ *of* $\bar{X}$ *the local states* $x_j = (x_{j1}, \dots, x_{n_j})$
are to be read as follows: If $n_j > 0$ *a customer of type* x_{jn_j} *is in service at node*
j, x_{jn_j-1} *is the type of the customer waiting at the head of the queue, ... , and*
x_{j1} *is the type of the customer who arrived most recently at node* j.

Proof: Let $p = (p(x, y) : x, y \in \tilde{S}(K, J))$ denote the transition matrix of X and
$\bar{p} = (\bar{p}(x, y) : x, y \in \tilde{S}(K, J))$ the transition matrix of $\bar{X}$. For any $x, y \in \tilde{S}(K, J)$
we have to show (see proposition 7.2 in the appendix.)

$$\pi^{K,J}(x)p(x, y) = \pi^{K,J}(y)\bar{p}(y, x). \tag{3.4}$$

Let for $x = (x_1, \dots, x_J) \in \tilde{S}(K; J)$, $x_j = (x_{j1}, \dots, x_{jn_j})$, $j = 1, \dots, J$ denote
$A(x) := \{j \in \{1, \dots, J\} : n_j > 0\}$ the set of busy nodes when the network
is in state x. One–step successor states $y = (y_1, \dots, y_J) \in \tilde{S}(K, J)$, $y_j = (y_{j1}, \dots, y_{jm_j}), j = 1, \dots, J$, of x, $y \neq x$, can be determined as follows and are
selected according to the following probabilities:

   (i) Chose some $B \subseteq A(x), B \neq \phi, B \neq \{1, \dots, J\}$, with probability

$$\prod_{j \in B} p_j(n_j) \cdot \prod_{j \in A(x)-B} q_j(n_j).$$

At nodes in B a service will expire at the end of the present time slot.

   (ii) For the customers of type x_{j1}, being in service at node j and now being
selected to leave that node, $j \in B$, chose their new types when arriving at node
$j + 1$ to be $y_{j+1,m_{j+1}}$ with probability

$$\prod_{j \in B} r(j; x_{j1}, y_{j+1,m_{j+1}}) \quad .$$

For $B \subseteq A(x), B \neq \phi, B \neq \{1, \dots, J\}$, define
$E(B) = \{j \in \{1, \dots, J\} : j \in B, j - 1 \notin B\}$,   (the END of B)
and
$F(B) = \{j \in \{1, \dots, J\} : j - 1 \in B, j \notin B\}$,   (in FRONT of B).

Then (3.4) with (3.2) reads

$$\left\{ \prod_{j=1}^{J} \left(\prod_{k=1}^{n_j} \eta(j; x_{jk}) \right) \left(\frac{\prod_{h=1}^{n_j-1} q_j(h)}{\prod_{h=1}^{n_j} p_j(h)} \right) \right\}$$

$$\cdot \prod_{j \in B} p_j(n_j) \cdot \prod_{j \in A(x)-B} q_j(n_j) \prod_{j \in B} r(j; x_{j1}, y_{j+1,m_{j+1}}) \qquad (3.5)$$

$$= \left\{ \prod_{j=1}^{J} \left(\prod_{k=1}^{m_j} \eta(j; y_{jk}) \right) \left(\frac{\prod_{h=1}^{m_j-1} q_j(h)}{\prod_{h=1}^{m_j} p_j(h)} \right) \right\} \cdot \prod_{j \in (B-E(B)) \cup F(B)} p_j(m_j)$$

$$\cdot \prod_{j \in (A(x)-B-F(B)) \cup E(B)} q_j(m_j) \cdot \prod_{j \in (B-E(B)) \cup F(B)} \bar{r}(j; y_{jm_j}, x_{j-1,1}).$$

First: Note that there is a bijection between those nodes being active in state x *and* showing a departure which leads in the original system from x to y and those nodes being active in state y *and* showing a departure which leads from y to x in the time reversed system. This bijection is given by:

$(B - E(B)) \cup F(B) = \{j \in \{1, \ldots, J\} : j - 1 \in B\}$ and
$B = \{j \in \{1, \ldots, J\} : j + 1 \in (B - E(B)) \cup F(B)\}$.

Second: From the definition of $\bar{r}(\cdot; \cdot, \cdot)$ it follows that for $j \in B$ (and therefore $j + 1 \in (B - E(B)) \cup F(B)$)

$$\eta(j; x_{j1}) \cdot r(j; x_{j1}, y_{j+1,m_{j+1}}) = \eta(j+1; y_{j+1,m_{j+1}}, x_{j1}) \cdot \bar{r}(j+1; y_{j+1,m_{j+1}}, x_{j1})$$

Third: Note that for $j \in E(B)$ $m_j = n_j - 1$, for $j \in F(B)$ $m_j = n_j + 1$, and for $j \in \{1, \ldots, J\} - F(B) - E(B)$ $m_j = n_j$.

Therefore (3.5) reduces to

$$\left\{ \prod_{j=1}^{J} \left(\frac{\prod_{h=1}^{n_j-1} q_j(h)}{\prod_{h=1}^{n_j} p_j(h)} \right) \right\} \cdot \prod_{j \in B} p_j(n_j) \cdot \prod_{j \in A(x)-B} q_j(n_j) \qquad (3.6)$$

$$= \left\{ \prod_{j \in F(B)} \left(\frac{\prod_{h=1}^{n_j} q_j(h)}{\prod_{h=1}^{n_j+1} p_j(h)} \right) \cdot \prod_{j \in E(B)} \left(\frac{\prod_{h=1}^{n_j-2} q_j(h)}{\prod_{h=1}^{n_j-1} p_j(h)} \right) \right.$$

$$\cdot \prod_{j \in \{1, \ldots, J\}-E(B)-F(B)} \left(\frac{\prod_{h=1}^{n_j-1} q_j(h)}{\prod_{h=1}^{n_j} p_j(h)} \right) \right\} \cdot$$

$$\cdot \left(\prod_{j \in F(B)} p_j(n_j + 1) \cdot \prod_{j \in B-E(B)} p_j(n_j) \right)$$

$$\cdot \left(\prod_{j \in E(B)} q_j(n_j - 1)^{\eta(1,n_j)} \cdot \prod_{j \in A(x)-B-F(B)} q_j(n_j) \right).$$

The rest of the proof is checking carefully (3.6). $\odot$

For obvious reasons the steady state distributions $\pi^{K,J}$ are said to be of *product form*, which merely reflects the similarity with the well-known continuous time results. In the closed network case this clearly does not imply independence of the steady state queue lengths. The structural difference between the equilibrium (3.2) and the continuous time analogue (see e.g. [75]), is clarified when considering the case of state independent service probabilities $p_j(n_j) = p_j$. Theorem 3.2 then simplifies to

Corollary 3.4 *Suppose we have $p_j(n_j) = p_j, n \in \mathbb{N}_+, j = 1, \ldots, J$. Then the unique stationary distribution of X is $\pi^{K,J} = (\pi^{K,J}(x) : x \in \tilde{S}(K,J))$ given by*

$$\pi^{K,J}(x_{11}, \ldots, x_{1n_1}; \ldots; x_{J1}, \ldots, x_{Jn_J}) \qquad (3.7)$$

$$= \prod_{j=1}^{J} \left(\prod_{k=1}^{n_j} \eta(j; x_{jk}) \right) \left(\frac{1}{q_j} \right)^{\eta(0,n_j)} \left(\frac{q_j}{p_j} \right)^{n_j} G(K,J)^{-1},$$

$$(x_{11}, \ldots, x_{Jn_J}) \in \tilde{S}(K,J).$$

It is the extra factor $\left(\frac{1}{q_j} \right)^{\eta(0,n_j)}$ for non empty queues which sets the difference. This factor e.g. is the reason why in the totally homogeneous cycle ($p_j = p, j = 1, \ldots, J$) the stationary distribution is not the uniform distribution on $\tilde{S}(K,J)$ as it is the case in continuous time. For conseqences of this observation see [106]. Note, that the extra factor $\left(\frac{1}{q_j} \right)^{\eta(0,n_j)}$ in (3.7) cannot be extracted in the general setting of theorem 3.2 because the missing factor $q_j(n)$ in the denominator there is not connected with (the appealing interpretation from (3.7)!) whether the node is empty or not, but is determined by the actual queue length.

The result of corollary 3.4 and even the case of state dependent service probabilities can in principle be proved by applying theorem 2.2 of [58] or by similar derivations as presented in section 4 of [8].

As pointed out in the introduction, we want to optimize systems with respect to throughput under constraints which are put on the behaviour of individual customers (cells). Possibly the more important problem might be the case of open systems. But it is well known from the continuous time setting that the closed case is important for such systems as well, e.g., when controlling open systems by window flow control mechanisms – see [110] and [111]. Further, from the results obtained in the closed case several results for open systems can be proved using limiting procedures. This is the case here as well. So we shall concentrate in the following on the computation of performance measures for the closed cycle of Bernoulli servers and turn in the next chapter to open systems.

To investigate the individual behaviour of customers we have to compute first the distribution of the network seen by a customer on his transit from one station to the next. It turns out that this arrival distribution is not the steady state of the system. This is clear in the closed cycle case but will occur in the open network case again. The latter is again in contrast to the continuous time analogue.

In continuous time a direct consequence of the product form equilibrium is the celebrated *Arrival Theorem*, [94], [123], which roughly states, that in equilibrium an arriving customer at node j observes the other customers being distributed according to the equilibrium of the same system if he himself would not be there, $j = 1, ..., J$. So the *arrival distribution* has the same structure as the equilibrium, and it is inpendent of the node j, where the customer arrives. (More formally: Interpreting the arrival distribution as Palm probabilities for product form queueing networks in equilibrium, we find that the Palm distribution is of the same product form structure as the equilibrium itself. See [86] for a discussion.) In our discrete time setting Palm distributions reduce to elementary conditional probabilities. For more details see the discussion after theorem 2.11.

Proposition 3.5 (Arrival Theorem) *Let* $X = (X(t) : t \in \mathbb{Z})$ *be the stationary continuation of* $X = (X(t) : t \in \mathbb{N})$ *under* $\pi^{K,J}$. *Assume that for node i and customer type m there exists some* m' *such that* $r(i - 1; m', m) > 0$, *i.e.* $m \in M(i)$, *and denote by* $A(i, m)$ *the event that at time 0 an arrival of a type-m customer at node i appeared,* $i \in \{1, ..., J\}$. *Then for* $x = (x_1, ..., x_J) \in \tilde{S}(K - 1, J)$

$$\pi_{i,m}^{K,J}(x_1, \ldots, x_J) \qquad (3.8)$$
$$:= P(X(0) = ((x_1, \ldots, x_{i-1}, (x_{i,1}, \ldots, x_{i,n_i}, m), x_{i+1}, \ldots, x_J)|A(i, m))$$
$$= \left(\prod_{k=1}^{n_i} \eta(i; x_{ik}) \right) \left(\frac{\prod_{h=1}^{n_i} q_i(h)}{\prod_{h=1}^{n_i} p_i(h)} \right)$$
$$\cdot \prod_{j=1, j\neq i}^{J} \left(\prod_{k=1}^{n_j} \eta(j; x_{jk}) \right) \left(\frac{\prod_{h=1}^{n_j-1} q_j(h)}{\prod_{h=1}^{n_j} p_j(h)} \right) \cdot G_{i,m}(K, J)^{-1}$$

where $G_{i,m}(K, J)^{-1}$ *is the norming constant.*

Proof: We abbreviate $X(t) =: X_t$ for $t \geq 0$, and consider the case $i = 1$ where

$$\pi_{1,m}^{K,J}(x_1, \ldots, x_J)$$
$$= P(\{X_0 = (x_{11}, \ldots, x_{1n_1}, m), x_2, \ldots, x_J)\} \cap A(1, m)) \cdot P(A(1, m))^{-1}.$$

Denote $((x_{11}, \ldots, x_{1n_1}, m), x_2, \ldots, x_J) =: \tilde{x}$, then

$$P(\{X_0 = (x_{11}, \ldots, x_{1n_1}, m), x_2, \ldots, x_J)\} \cap A(1, m))$$
$$= \sum_{y \in S(K,J)} P(\{X_0 = (x_{11}, \ldots, x_{1n_1}, m), x_2, \ldots, x_J)\} \cap \{X_{-1} = y\} \cap A(1, m))$$
$$= \sum_{y \in S(K,J)} 1_{[\{(X_{-1}, X_0) = (y, \tilde{x})\} \subseteq A(1, m)]} P(X_{-1} = y, X_0 = \tilde{x})$$
$$= \sum_{y \in S(K,J)} 1_{[\{(X_{-1}, X_0) = (y, \tilde{x})\} \subseteq A(1, m)]} P(X_0 = \tilde{x}) \cdot P(X_{-1} = y|X_0 = \tilde{x})$$

$$\overbrace{}^{=p_1(n_1+1)}$$

$$= \ \pi^{K,J}(\tilde{x}) \cdot \sum_{y \in S(K,J)} 1_{[\{(X_{-1},X_0)=(y,\tilde{x})\} \subseteq A(1,m)]} \bar{p}(\tilde{x},y)$$

$$= \ \left(\prod_{k=1}^{n_1} \eta(1; x_{1k})\right) \left(\frac{\prod_{h=1}^{(n_1+1)-1} q_1(h)}{\prod_{h=1}^{n_1+1} p_1(h)}\right) \cdot \eta(1; m)$$

$$\cdot \prod_{j=2}^{J} \left(\prod_{k=1}^{n_j} \eta(j; x_{jk})\right) \left(\frac{\prod_{h=1}^{n_j-1} q_j(h)}{\prod_{h=1}^{n_j} p_j(h)}\right) \cdot G(K,J)^{-1} \cdot p_1(n_1+1).$$

It follows

$$\pi_{1,m}^{K,J}(x_1, \ldots, x_J)$$

$$= \ \left(\prod_{k=1}^{n_1} \eta(1; x_{1k})\right) \left(\frac{\prod_{h=1}^{n_1} q_1(h)}{\prod_{h=1}^{n_1} p_1(h)}\right) \cdot \eta(1; m)$$

$$\cdot \prod_{j=2}^{J} \left(\prod_{k=1}^{n_j} \eta(j; x_{jk})\right) \left(\frac{\prod_{h=1}^{n_j-1} q_j(h)}{\prod_{h=1}^{n_j} p_j(h)}\right) \cdot G(K,J)^{-1} \cdot P(A(1;m))^{-1}$$

and

$$G_{1,m}(K,J)^{-1} = G(K,J)^{-1} \cdot \frac{\eta(1; m)}{P(A(1;m))}$$

is the norming constant. $\odot$

A different form of an arrival theorem was proved by Henderson and Taylor ([58]): They computed in a rather general setting the disposition probability for stay–on customers seen by a prescribed set of departing customers just before the latter enter their destination node. These probabilities can be computed in a way similar to the proof of theorem 3.5.
And the other way round: Having the arrival probabilities of [58], Corollary 2.4, at hand, by a suitable, but lengthy, summation our result specialized to the case of indistinguishable customer would follow.
The elementary proof given for the case of networks with different customer types is from [27].

Remark 3.6 *We extend our definitions to include computations concerning different population sizes:*
For any $k \in I\!N$ $\tilde{S}(k, J)$ is the set of configurations which a population of k customers with possible types M at the J nodes may show. (Recall the Assumption 3.1 and the remarks there. These apply here again.)
The remarks on the steady state and arrival probablities can be summarized:
$\pi_{i,m}^{K,J}$ is defined on $\tilde{S}(K-1, J)$ and $\pi_{i,m}^{K,J} \neq \pi^{K-1,J}$ holds in contrast to the continuous time case where $\pi_{i,m}^{K,J} = \pi^{K-1,J}$ holds. Further in continuous time we have $\pi_{i,m}^{K,J} = \pi_{j,m}^{K,J}$, while in our discrete time cycles $\pi_{i,m}^{K,J} \neq \pi_{j,m}^{K,J}$ holds.
Some further details of the connection between norming constants in different

systems of similar structure can be found in [136], pp. 144,145. It is shown there that a fixed sequencing of customers in a cycle without the possibility of overtaking causes even in continuous time $\pi_{i,m}^{K,J} \neq \pi^{K-1,J}$, where again the LHS is to be read as the steady state in a system without customer m. This phenomenon occurs here as well.

3.2 Closed Cycles of Bernoulli Servers – The Multichain Case

Many applications in computer and computer communication performance evaluation require to distinguish between different classes of customers (jobs). Jobs of a specified class (*chain*) are indistinguishable with respect to their behaviour in the system but may individually show a versatile behaviour as described in section 2.2. For a short introduction into the models and the necessary techniques for continuous time systems see [37]. We describe the modeling principles in the framework of cyclic queues and point out the differences arising compared with the unichain case.

Consider a closed cycle of state–dependent Bernoulli servers as described in Section 3.1 where customers belong to different chains which define their behaviour with respect to the types. The set of customers is split into k disjoint subsets $\mathcal{K}_g$ with $|\mathcal{K}_g| = K_g \in \mathbb{N}_+, g = 1, \ldots, k, \sum_{g=1}^{k} K_g = K$. Each set $\mathcal{K}_g$ constitutes a specified chain of customers which are undistinguishable within each chain. So each set $\mathcal{K}_g$ is associated to a subset $M_g \subseteq M$ of types, where the M_g are pairwise disjoint with $\sum_{g=1}^{k} M_g = M$. No customer ever leaves his associated set $\mathcal{K}_g$ or set M_g. For the matrices $r(i)$ it follows:

$$r(i; m, m') > 0 \quad only \quad if \quad m, m' \in M_g \quad for \quad some \quad g \in \{1, \ldots, k\}.$$

We now assume that for each set $\mathcal{K}_g, g \in \{1, \ldots, k\}$ the system of equations

$$\eta(i; m) = \sum_{m' \in M_g} \eta(i-1; m')r(i-1; m', m), \quad i = 1, \ldots, J, \ m \in M_g,$$

has a unique stochastic solution $\eta = (\eta(i; m) : i = 1, \ldots, J, \ m \in M_g)$. (Note that type m determines g, $\mathcal{K}_g$, and M_g uniquely.)
The state space of the describing process $X = (X(t) : t \geq 0)$ is denoted again by

$$\tilde{S} := \tilde{S}(K, J),$$

which is now defined by

$$\{(x_1, \ldots, x_J) \in \tilde{S}(K, J) \Longleftrightarrow$$
$$x_j = (x_{j1}, \ldots, x_{jn_j}) \in M(j)^{n_j}, j = 1, \ldots, J,$$
$$and \quad \sum_{j=1}^{J} \sum_{h=1}^{n_j} 1\{x_{jh} \in \mathcal{K}_g\} = k_g, \quad \forall g = 1, \ldots, k.$$

In the multichain case we have a complete analogue of Theorem 3.2 and Proposition 3.3, which especially yields (remember the Assumption 3.1) for the steady state probabilities that they are of the form (3.2), with $\tilde{S}(K, J)$ as described here. The reformulation of the arrival theorem for the multichain case is slightly more involved, but using the notations of Section 3.1 with a more elaborated interpretation yields formally (3.8) again. The problem is: Prescribing for some node i and type m the event $A(i, m) \in \mathcal{F}$, that at time 0 a type–m arrival at node i appeares, we have fixed the chain, say $\mathcal{K}_f$, where the arriving customer belongs to, $f \in \{1, \ldots, k\}$. For different chains the possible dispositions of the other customers seen by the arrivals will be different and

$$\pi_{i,m}^{K,J}(x_1, \ldots, x_J)$$
$$:= \quad P(X(0) = ((x_1, \ldots, x_{i-1}, (x_{i,1}, \ldots, x_{i,n_i}, m), x_{i+1}, \ldots, x_J)|A(i, m))$$

is now defined for

$$(x_1, \ldots, x_J) \in \tilde{S}_f(K, J)$$

only, where

$$\{(x_1, \ldots, x_J) \in \tilde{S}_f(K, J) \Longleftrightarrow$$
$$x_j = (x_{j1}, \ldots, x_{jn_j}) \in M^{n_j}(j), j = 1, \ldots, J,$$
$$and \quad \sum_{j=1}^{J}\sum_{h=1}^{n_j} 1\{x_{jh} \in \mathcal{K}_g\} = k_g, \quad \forall g \in \{1, \ldots, k\} - \{f\}$$
$$and \quad \sum_{j=1}^{J}\sum_{h=1}^{n_j} 1\{x_{jh} \in \mathcal{K}_f\} = k_f - 1.$$

In the rest of the notes we shall restrict our attention to the unichain case. The way the results carry over to the multichain case will always be obvious. Adapting the closed cyle and the open tandem performance evaluation to the multichain case is then direct by using the techniques of the previous sections.

3.3 Delay Time Distribution for Customers in a Closed Cycle of State Independent Bernoulli Servers

For a closed cycle of J state independent Bernoulli servers under FCFS queueing regime with unlimited waiting room as described in section 3.1 and specified in corollary 3.4 we derive the steady state cycle time for a customer in the system, and similar quantities. Most important is to determine the joint distribution of successive sojourn times of a customer at the different nodes during such a cycle.

From the arrival theorem 3.5 and the type independent service times it follows that these distributions do not depend on the type of the cycling customer. We therefore can and will restrict our attention to the case of indistinguishable customers. For ease of reference we state the relevant steady state results for this specific case before proceding.

A Markovian description of the time evolution of the system is provided by the joint queue length process $X = (X(t) : t \in \mathbb{N})$ with state space

$$S(K, J) = \{(x_1, \ldots, x_J) \in \mathbb{N}^J : x_1 + \ldots + x_J = K\}.$$

X records the development of the joint queue length vector, i.e., $X(t) = (X_1(t), \ldots, X_J(t)) = (x_1, \ldots, x_J)$ indicates that at time t there are x_j customers present at node j, including the one in service, if any, $j = 1, \ldots, J$. For the Markov chain X the following corollaries specialize the results of section 3.1.

Theorem 3.7 *([107]) $X = (X(t) : t \in \mathbb{N})$ is ergodic on $S(K, J)$ with unique limiting and steady-state distribution $\pi^{K,J} = (\pi^{K,J}(x) : x \in S(K, J))$ given by*

$$\pi^{K,J}(x_1, \ldots, x_J) \quad = \quad \prod_{j=1}^{J} \left(\frac{1}{q_j}\right)^{\eta(0, x_j)} \left(\frac{q_j}{p_j}\right)^{x_j} \cdot G(K, J)^{-1}, \qquad (3.9)$$

$$(x_1, \ldots, x_J) \in S(K, J),$$

where $\eta(a, b) = \begin{cases} 1 & if \quad a \neq b \\ 0 & if \quad a = b \end{cases}$, and the norming constant is

$$G(K, J) \quad = \quad \sum_{(y_1, \ldots, y_J) \in S(K, J)} \prod_{j=1}^{J} \left(\frac{1}{q_j}\right)^{\eta(0, y_j)} \left(\frac{q_j}{p_j}\right)^{y_j}.$$

Proposition 3.8 *Let $X = (X(t) : t \in \mathbb{Z})$ be the stationary continuation of $(X(t) : t \in \mathbb{N})$ under $\pi^{K,J}$, and $\bar{X} = (\bar{X}(t) : t \in \mathbb{Z})$ its time reversal, defined pathwise by*

$$\bar{X}(t, \omega) = X(-t, \omega), t \in \mathbb{Z}, \qquad \omega \in \Omega.$$

Then $\bar{X}$ is the steady-state joint queue length process of a cycle of queues $1, 2, \ldots, J$, with Bernoulli servers under FCFS. The parameter of the geometrically distributed service times at node j is $p_j, j = 1, \ldots, J$. A customer leaving node j proceeds to node $j - 1$, where 1 - 1 := J. Shortly:

The time reversal $\bar{X}$ of X describes the evolution of the system when the direction of customer movements is reversed.

$\pi^{K,J}$ is said to be of *product form* ([107]), resembling the specification of the analogous steady-state probabilities of continuous time systems, see [75]. In fact their result is stated in a slightly different, but equivalent form: Our extra factor $\left(\frac{1}{q_j}\right)^{\eta(0, x_j)}$ for non-empty nodes is substituted by an extra factor $q_j^{\delta(0, x_j)}$ for empty nodes, and the norming constant is changed accordingly.

Theorem 3.9 *Let $X = (X(t) : t \in \mathbb{Z})$ be the stationary continuation of $(X(t) : t \in I\!N)$ under $\pi^{K,J}$ and denote by $A(i)$ the event that at time 0 an arrival at appeared at node i, $i \in \{1, \dots, J\}$.*
Then for $(x_1, \dots, x_J)$ with $x_1 + \dots + x_J = K - 1$

$$\pi_i^{K,J}(x_1, \dots, x_J) \tag{3.10}$$
$$:= P(X(0) = (x_1, \dots, x_{i-1}, x_i + 1, x_{i+1}, \dots, x_J) \mid A(i))$$
$$= \left(\frac{q_i}{p_i}\right)^{x_i} \prod_{j=1, j \neq i}^{J} \left(\frac{1}{q_j}\right)^{\eta(0, x_j)} \left(\frac{q_j}{p_j}\right)^{x_j} \cdot G_i(K, J)^{-1}$$

where $G_i(K, J)$ is the norming constant.

Remark 3.10 *(a) Similarly to remark 3.6 the arrival probabilities in discrete time cyclic queues depend on where there arrival appears: (3.10) is asymmetric with respect to the node number i, and the arrival probabilities keep trace of the arriving customer – which is not the case in continuous time.*
(b) Extending our definitions to general finite population sizes and node sets we shall use expressions like $S(K - k, J - 1), \pi^{K-1,J}, \pi_i^{K-1,J}$, etc., the meaning of which will be clear from the context. In continuous time we then have on $S(K - 1, J)$: $\pi^{K-1,J} = \pi_i^{K,J}, i = 1, \dots, J$, ([123] and [94]) while in discrete time we have $\pi_i^{K,J} \neq \pi^{K-1,J}$, and $\pi_i^{K,J} \neq \pi_j^{K,J}$ for $i \neq j, i, j = 1, \dots, J$.

We now consider a test customer C_0 arriving at time 0 at node 1 who finds the other customers distributed according to $\pi_1^{K,J}$. Let $(S_1, \dots, S_J)$ denote the vector of C_0's successive sojourn times (=waiting time + service time) at the nodes during his first cycle starting at time 0.

We denote by $P_{\pi_1^{K,J}}$ a probability law on $(\Omega, \mathcal{A})$ which governs the evolution of the system under this condition, and by $E_{\pi_1^{K,J}}[\cdot]$ expectations under $P_{\pi_1^{K,J}}$. We shall prove the following theorem [29]:

Theorem 3.11 *The joint distribution of the vector $(S_1, \dots, S_J)$ is given by its generating function*

$$E_{\pi_1^{K,J}}\left[\prod_{j=1}^{J} \theta_j^{S_j}\right] \tag{3.11}$$

$$= \sum_{(x_1, \dots, x_J) \in S(K-1, J)} G_1(K, J)^{-1} \left(\frac{q_1}{p_1}\right)^{x_1} \prod_{j=2}^{J} \left\{\left(\frac{1}{q_j}\right)^{\eta(0, x_j)} \left(\frac{q_j}{p_j}\right)^{x_j}\right\}$$

$$\cdot \left(\frac{p_1 \theta_1}{1 - q_1 \theta_1}\right)^{x_1+1} \prod_{j=2}^{J} \left\{\left(\frac{p_j \theta_j}{1 - q_j \theta_j}\right)^{x_j+1} \left(\frac{1}{\theta_j}\right)^{\eta(0, x_j)}\right\},$$

$$|\theta_j| \leq 1, j = 1, \dots, J.$$

It is important that the appealing interpretation of the RHS of (3.11) as a direct result of conditioning on the arrival situation is false:

$$\left(\frac{p_1\theta_1}{1-q_1\theta_1}\right)^{x_1+1}\prod_{j=2}^{J}\left\{\left(\frac{p_j\theta_j}{1-q_j\theta_j}\right)^{x_j+1}\left(\frac{1}{\theta_j}\right)^{\eta(0,x_j)}\right\}$$

is **not** the joint conditional joint sojourn time distribution $\mathcal{L}(S_1,\dots,S_J\mid X(0)=(x_1+1,x_2,\dots,x_J))$. The proof of this can be done by inserting these expressions into the set of first entrance equations for the conditional sojourn time LST in (3.21) and then verifying that they do not solve the system.

Interchanging the numbering of the nodes we have an immediate result for other cycles. Note that in the transform formula (3.11) the asymmetry of (3.10) formally reappears, but compare this with the statement in Corollary 3.15 below.

Corollary 3.12 *If C_0 arrives at time 0 at node $i \in \{1,\dots,J\}$, finding the other customers distributed according to $\pi_i^{K,J}$, and if $(S_1^{(i)}, S_2^{(i)},\dots, S_J^{(i)})$ denotes the vector of his sojourn times during his cycle which starts at 0, then*

$$E_{\pi_i^{K,J}}\left[\prod_{j=1}^{J}\theta_j^{S_j^{(i)}}\right]\tag{3.12}$$

$$=\sum_{(x_1,\dots,x_j)\in S(K-1,J)}G_i(K,J)^{-1}\left\{\prod_{j=1}^{J}\left(\frac{p_j\theta_j}{1-q_j\theta_j}\right)\right\}$$

$$\cdot\left(\frac{q_i\theta_i}{1-q_i\theta_i}\right)^{x_i}\prod_{j=1,j\neq i}^{J}\left\{\left(\frac{1}{q_j\theta_j}\right)^{\eta(0,x_j)}\left(\frac{q_j\theta_j}{1-q_j\theta_j}\right)^{x_j}\right\},$$

$$|\theta_j|\le 1, j=1,\dots,J.$$

Proof of Theorem 3.11: The proof is by induction on the length of the cycle and the number of customers. We consider the cycle built of an initial node connected to a smaller residual cycle with less customers cycling, where the induction hypothesis applies.

We describe the dynamics of the network during C_0's first cycle after 0, recording carefully the actual position of C_0 relative to the other customers. A sufficiently detailed description of the system will be provided by using the following notation:

Let for $i \in \{1,\dots,J\}$ be $\tilde{x}_i, x_1, x_2,\dots,x_J \in \mathbb{N}$ such that $\tilde{x}_i + \sum_{j=1}^{J}x_j = K-1$. Then saying the network is in *state*

$$(x_1,\dots,x_{i-1},\tilde{x}_i\mid x_i,x_{i+1},\dots,x_J)$$

means that there are x_j customers present at nodes $j\in\{1,\dots,J\}-\{i\}, C_0$ is present at node i, seeing exactly x_i other customers before and $\tilde{x}_i$ other customers behind him.

We denote the set of all such possible states, where C_0 is represented by the vertical bar at some node $i, i = 1,\dots,J$, by $S(C_0)$. Using this notation, the possible *arrival states* for C_0 at node 2 are of the form $(x_1,0\mid x_2,x_3,\dots,x_J)$, where $(x_1,x_2,\dots,x_J)\in S(K-1,J)$.

The dynamics of the system during C_0's first cycle is described by a Markov chain $Z = (Z(t) : t \geq 0)$ on the state space $S(C_0)$. The transition law and the behaviour of Z are coupled with that of X in an obvious way. We therefore use similar notation for quantities connected with these processes. E.g., we define Z on an underlying probability space $(\Omega, \mathcal{A}, P_{\pi_1^{K,J}})$, which means

$$P_{\pi_1^{K,J}}(Z(0) = (0 \mid x_1, x_2, \dots, x_J)) = \pi_1^{K,J}(x_1, x_2, \dots, x_J), \qquad (3.13)$$

$$(x_1, x_2, \dots, x_J) \in S(K-1, J).$$

It should be noted that Z under $P_{\pi_1^{K,J}}$ is not stationary; because $P_{\pi_1^{K,J}}$ is a probability measure we conclude from (3.13) for $\tilde{x}_1 > 0$

$$P_{\pi_1^{K,J}}(Z(0) = (\tilde{x}_1 \mid x_1, x_2, \dots, x_J)) = 0, \quad \forall (x_1, x_2, \dots, x_J) \in S(K - \tilde{x}_1 - 1, J).$$

We denote the 1-step transition law of Z by $p^Z = (p^Z[x, y] : x, y \in S(C_0))$. Our first step in analyzing the distribution of $(S_1, \dots, S_J)$ is applying the total probability formula to discriminate between different arrival dispositions seen by C_0 on his arrival at node 2:

$$E_{\pi_1^{K,J}} \left[\prod_{j=1}^{J} \theta_j^{S_j} \right] \qquad (3.14)$$

$$= \sum_{(x_1, \dots, x_J) \in S(K-1, J)} \sum_{t_1 = 1}^{\infty} P_{\pi_1^{K,J}}(S_1 = t_1, Z(S_1) = (x_1, 0 \mid x_2, x_3, \dots, x_J)) \theta_1^{t_1}$$

$$\cdot \ E_{\pi_1^{K,J}} \left[\prod_{j=2}^{J} \theta_j^{S_j} \mid S_1 = t_1, Z(S_1) = (x_1, 0 \mid x_2, x_3, \dots, x_J) \right].$$

The key step of our proof is the following lemma the proof of which will be postponed to the end of this section.

Lemma 3.13 *The joint distribution of $(S_1, Z(S_1))$ under $P_{\pi_1^{K,J}}$ is given by*

$$\sum_{t_1 = 1}^{\infty} P_{\pi_1^{K,J}}(S_1 = t_1, Z(S_1) = (x_1, 0 \mid x_2, x_3, \dots, x_J)) \theta_1^{t_1} \qquad (3.15)$$

$$= \pi_2^{K,J}(x_1, x_2, \dots, x_J) \left(\frac{p_1 \theta}{1 - q_1 \theta} \right)^{x_1 + 1} \left(\frac{1}{\theta} \right)^{\eta(0, x_1)}.$$

The strong Markov property of Z ensures that $(S_2, \dots, S_J)$ is conditionally independent of S_1 given $Z(S_1) = (x_1, 0 \mid x_2, x_3, \dots, x_J)$. Using this property and inserting (3.15) into (3.14) yields

$$E_{\pi_1^{K,J}} \left[\prod_{J=1}^{J} \theta_j^{S_j} \right] \qquad (3.16)$$

$$= \sum_{(x_1, x_2, \dots, x_J) \in S(K-1, J)} \pi_2^{K,J}(x_1, x_2, \dots, x_J) \left(\frac{p_1 \theta_1}{1 - q_1 \theta_1} \right)^{x_1 + 1} \left(\frac{1}{\theta_1} \right)^{\eta(0, x_1)}$$

$$\cdot \quad E\left[\prod_{j=2}^{J} \theta_j^{S_j} \mid Z(0) = (x_1, 0 \mid x_2, x_3, \ldots, x_J)\right]$$

$$= \quad G_2(K, J)^{-1} \sum_{x_1=0}^{K-1} \left\{\left(\frac{1}{q_1\theta_1}\right)^{\eta(0,x_1)} \left(\frac{q_1}{p_1}\right)^{x_1}\right\} \left(\frac{p_1\theta_1}{1 - q_1\theta_1}\right)^{x_1+1}$$

$$\cdot \quad G_2(K - x_1, J - 1) \sum_{(x_2, x_3, \ldots, x_J) \in S(K-1-x_1, J-1)} G_2(K - x_1, J - 1)^{-1} \left(\frac{q_2}{p_2}\right)^{x_2}$$

$$\cdot \quad \prod_{j=3}^{J} \left\{\left(\frac{1}{q_j}\right)^{\eta(0,x_j)} \left(\frac{q_j}{p_j}\right)^{x_j}\right\} E\left[\prod_{j=2}^{J} \theta_j^{S_j} \mid Z(0) = (x_1, 0 \mid x_2, x_3, \ldots, x_J)\right].$$

We now use the overtake-free property of the closed tandem system under consideration; during one cycle of C_0 customers who jump (from node J) to some node behind C_0 can no longer influence C_0's future sojourn at the nodes of his residual cycle. This justifies the following equation:
For $(x_2, x_3, \ldots, x_J) \in S(K - x_1 - 1, J - 1)$

$$E\left[\prod_{j=2}^{J} \theta_j^{S_j} \mid Z(0) = (x_1, 0 \mid x_2, x_3, \ldots, x_J)\right] \tag{3.17}$$

$$= \quad E\left[\prod_{j=2}^{J} \theta_j^{S_j} \mid Z(0) = (0 \mid x_2, x_3, \ldots, x_J)\right],$$

where the RHS is to be interpreted as the conditional joint sojourn time of a test customer C_0 in a cyclic tandem of nodes $2, 3, \ldots, J$, given C_0 finds an arrival disposition $(0 \mid x_2, x_3, \ldots, x_J)$ of other customers.

Using induction with respect to the number of nodes and the number of customers cycling we transform (3.16) into

$$E_{\pi_1^{K,J}}\left[\prod_{j=1}^{J} \theta_j^{S_j}\right] \tag{3.18}$$

$$= \quad \sum_{(x_1, x_2, \ldots, x_J) \in S(K-1, J)} \pi_2^{K,J}(x_1, x_2, \ldots, x_J)$$

$$\cdot \quad \left(\frac{p_2\theta_2}{1 - q_2\theta_2}\right)^{x_2+1} \prod_{j=1, j\neq 2}^{J} \left(\frac{p_j\theta_j}{1 - q_j\theta_j}\right)^{x_j+1} \left(\frac{1}{\theta_j}\right)^{\eta(0,x_j)}.$$

That the expectation in the LHS of (3.18) is expressed as an expectation under $P_{\pi_2^{K,J}}$ in the RHS resembles the similar property in the continuous time setting. (See [86] for details, or [30].) There an invariance property of the joint sojourn time distribution under exchanging the underlying measure can be proved a priori. ([76], Theorem 3.1.) Here this must be proved as a final step of our proof of theorem 3.11. But the RHS of (3.18), which is (3.12) for $i = 2$, is a symmetric function by the lemma 7.3 of Appendix 7.2. The proof is completed.⊙

Remark 3.14 *The proof of theorem 3.11 is performed by induction on the length of the cycle and the number of customers. We considered the cycle built of an initial node connected to a smaller residual cycle with less customers cycling, where the induction hypothesis applies. Insofar the proof is standard, mimicing the continuous time analogue, see [10], [76]. Kelly and Pollett showed even more: In the continuous time setting the splitting of the cycle for the induction step is possible between every pair of nodes j and $j + 1$, $j \in \{1, 2, \ldots, J - 1\}$ – the induction step is always the same ! The main difficulty that arises in the discrete time system is to justify this splitting. Directly carrying over the arguments from [76] is not possible, because of the dependence of the disposition distribution for the other customers on the node where the customer jumps (theorem 3.9 and remark 3.10).*

Following the ideas and the lines of the proof in [76], we would obtain results on the distribution of $(S_1, \ldots, S_J)$ for a cycling customer in systems governed by measures $P_{\pi_j^{K,J}}, j = 2, \ldots, J$, but not the required result for a system governed by $P_{\pi_1^{K,J}}$.

In continuous time it can be proved a priori (without computing explicit distributions !) that for the joint distributions $P_{\pi_1^{K,J}}^{(S_1, \ldots, S_J)} =: P_{\pi_1^{K,J}}^{(S_1^{(1)}, \ldots, S_J^{(1)})} = P_{\pi_j^{K,J}}^{(S_1^{(j)}, \ldots, S_J^{(j)})}$ holds, $j = 2, \ldots, J$ (we use the notation introduced above for similar quantities). This is Theorem 3.1 [76] and is used in the sojourn time Theorem 6.1 (p.650) there. For further detailes concerning these structural properties of continuous time network processes see [86] or [35].

Following the remark 3.14, our theorem 3.9 suggests that a formal analogue of Theorem 3.1 of [76] should not hold in discrete time. The surprising conclusion from theorem 3.11 and corollary 3.12 is that in case of sojourn time distributions an analogue of the symmetry properties of continuous time systems derived in Theorem 3.1 of [76] does hold in discrete time. Indeed: Similarly to the last step in the proof of Theorem 3.11 we conclude from Corollary 3.12 and Lemma 7.3 the following result.

Corollary 3.15 *Let C_0 commence at time 0 a cycle starting at any node $i \in \{1, \ldots, J\}$, seeing at this time instant the other customers distributed according $\pi_i^{K,J}$. Then the joint distribution of C_0's successive sojourn times vector $(S_1, \ldots, S_J)$ during this cycle does not depend on the node i, where the cycle is started.*

As a by-product of Theorem 3.11 and Corollary 3.15 we have a result on the distribution of cycle times (= sum of successive sojourn times) which is the end–to–end–delay e.g. in a transmission line under window flow control for a system in heavy traffic (see [110] and [111]). A direct proof of different character for the result on the cycle time distribution is given in [26].

Corollary 3.16 (End–to–end–delay) *In the setting of Corollary 3.12 the cycle time distribution of C_0 starting at node i is given by its generating function*

$$E_{\pi_i^{K,J}}\left[\theta^{S_1^{(i)}+\cdots+S_J^{(i)}}\right] \tag{3.19}$$

$$= \sum_{(x_1,\ldots,x_J)\in S(K-1,J)} G_i(K,J)^{-1} \left\{\prod_{j=1}^{J} \frac{p_j\theta}{1-q_j\theta}\right\}$$

$$\cdot \left(\frac{q_i\theta}{1-p_i\theta}\right)^{x_i} \prod_{j=1,j\neq i}^{J} \left\{\left(\frac{1}{q_j\theta}\right)^{\eta(0,x_j)}\left(\frac{q_j\theta}{1-q_j\theta}\right)^{x_j}\right\},$$

$$|\theta|\leq 1, \quad i=1,\ldots,J.$$

This distribution does not depend on the node where the cycle starts.

The results of Theorem 3.11 and Corollary 3.15 provide us with the full information about the distribution of the vector of sojourn times. To consider only parts of the cycle is simply to compute marginal distributions of any order. Especially moments and covariances can be computed explicitly. Explicit results on partial-cycle times are given in [26], section 4. Similar results for joint sojourn time distributions can be proved similarly.

It remains to prove Lemma 3.13. This is indeed the core of our analysis because it elaborates in detail on the structure put on our system by the discrete time scale. It provides us with a substitute (of completely different nature) of the mentioned Theorem 3.1 in [76]. For more details see [29].

Proof of Lemma 3.13: The skeleton of the proof is as follows:
We show that a system of *quasi–first–entrance–equations* ((3.21) below) has nearly a *product form solution*. The combinatorial complexity of the system (due to the appearance of *multiple events* in the running system) prevents us from directly solving this system of equations. Instead in the course of an inductive procedure, we identify certain substructures which allow a probabilistic interpretation in terms of related time reversed systems. Applying the operation of time reversal to **these** systems we are able to transform the original equations into a form amenable to a solution.
In our proof of Lemma 3.13 we will **not** elaborate on the pathwise time reversed joint queue length process $\bar{X}$, as was done in the continuous time setting by e.g. [10] or [86]. Our intuitive arguments evoking time reversed systems of different population sizes only describe (from Proposition 3.8) which sums are to be balanced, using

$$\pi^{k,J}(x)\cdot p(x,y) = \pi^{k,J}(y)\cdot \bar{p}(y,x), x,y \in S(k,J),$$

where $p(\cdot,\cdot)$ is the transition kernel of X, and $\bar{p}(\cdot,\cdot)$ that of $\bar{X}$. We therefore are able to avoid arguing with (stationary) probabilities in a non–stationary system - the latter turns into a non-homogeneous Markov system under time reversal.
A second point why we could not start with considering the system at a departure instant from node 1 and then looking forward and backward in times

is the dependence of the arrival distribution on the specific arrival node. So we cannot prove a fortiori that the distribution of $(S_1.S_2, \dots, S_J)$ is the same under any P_{π_j}.

Our main work is therefore a careful exposition of the equations concerning subsystems, which always occur in connection with some *nuisance terms* that have to be extracted.

For notational convenience we abbreviate $P := P_{\pi_1^{K,J}}$. We have for $(x_1, x_2, x_3, \dots, x_J) \in S(K-1, J)$ and $t \geq 1$

$$P(S_1 = t, Z(S_1) = (x_1, 0 \mid x_2, x_3, \dots, x_J)) \qquad (3.20)$$

$$= \sum_{y \in S(C_0)} P(S_1 > t - 1, Z(t-1) = y) \cdot p^Z[y, (x_1, 0 \mid x_2, x_3, \dots, x_J)]$$

$$= \sum_{(y_1 \mid 0, y_2, y_3, \dots, y_J) \in S(C_0)} P(Z(t-1) = (y_1 \mid 0, y_2, y_3, \dots, y_J)$$

$$\cdot p^Z[(y_1 \mid 0, y_2, y_3, \dots, y_J), (x_1, 0 \mid x_2, x_3, \dots, x_J)].$$

In a first step we compute these transient state probabilities. It is possible to reduce the sojourn time problem for S_1 to determining the transient state probabilities of Z. We may restrict our attention on $Z's$ development conditioned on being in

$$S_1(C_0) = \{(\tilde{x}_1 \mid x_1, x_2, \dots, x_J) : \tilde{x}_1 + x_1 + x_2 + \dots + x_J = K - 1\} \subseteq S(C_0).$$

Denote for $x \in S_1(C_0)$

$$f[x; \theta] := \sum_{t=0}^{\infty} \theta^t \cdot P(Z(t) = x), \quad \mid \theta \mid \leq 1.$$

These generating functions are uniquely defined by the following set of *quasi–first–entrance–equations*.

For $x = (\tilde{x}_1 \mid x_1, x_2, \dots, x_J) \in S_1(C_0)$

$$f[x; \theta] - \pi_1^{K,J}(x_1, x_2, \dots, x_J) \cdot \delta(\tilde{x}_1, 0) = \theta \cdot \sum_{y \in S_1(C_0)} f[y; \theta] \cdot p^Z[y, x]. \qquad (3.21)$$

($\delta(\cdot, \cdot)$ is Kronecker's Delta.) We shall prove that the solution of (3.21) is

$$f[(\tilde{x}_1 \mid x_1, x_2, \dots, x_J); \theta] \qquad (3.22)$$

$$= \pi_1^{K,J}(\tilde{x}_1 + x_1, x_2, \dots, x_J) \cdot \left(\frac{p_1 \theta}{1 - q_1 \theta}\right)^{\tilde{x}_1} \cdot \frac{1}{1 - q_1 \theta},$$

$$(x_1, x_2, \dots, x_J) \in S(K - 1 - \tilde{x}_1, J), \quad \mid \theta \mid \leq 1,$$

Exploiting the form of $\pi_1^{K,J}$ in (3.10) of Theorem 3.9 we see that the solution of (3.21) has nearly a *product form*. (3.22) will be proved by induction on $\tilde{x}_1$.

Let us first consider $p^Z[\cdot,\cdot]$ in detail. For $(\tilde{x}_1 \mid x_1, x_2, \ldots, x_J) =: x$ let $A(x) = \{j \in \{1, \ldots, J\} : x_j > 0\}$ be the set of busy nodes, where *other customers* (not C_0) are in service. Then a transition of Z on $S_1(C_0)$ is uniquely determined by the set $B \subseteq A(x)$ of those nodes where the service of customers expires.

For any $x \in S_1(C_0)$ and any $B \subseteq A(x)$ we define $y := H_B(x) \in S_1(C_0)$ to be the successor state of x obtained by transfering the customers in service at nodes in B to the next node. This happens with probability

$$p^Z[x, H_B(x)] = \left(\prod_{j \in B} p_j\right) \cdot \left(\prod_{j \in A(x)-B} q_j\right) \cdot q_1^{\delta(0,x_1)}. \qquad (3.23)$$

Here the last factor counts for the case that C_0 is in service at node 1, but is not allowed to leave that node, because Z should stay in $S_1(C_0)$. Now for $\tilde{x}_1 = 0$ inserting (3.22) into (3.21) yields

$$\left(\frac{q_1}{p_1}\right)^{x_1} \cdot \prod_{j=2}^{J} \left\{\left(\frac{1}{q_j}\right)^{\eta(0,x_j)} \left(\frac{q_j}{p_j}\right)^{x_j}\right\} \cdot \left[\frac{1}{1-q_1\theta} - 1\right] \qquad (3.24)$$

$$= \theta \cdot \sum_{(0|y_1,y_2,\ldots,y_J) \in S_1(C_0)} \left(\frac{q_1}{p_1}\right)^{y_1} \cdot \prod_{j=2}^{J} \left\{\left(\frac{1}{q_j}\right)^{\eta(0,y_j)} \cdot \left(\frac{q_j}{p_j}\right)^{y_j}\right\} \cdot \left(\frac{1}{1-q_1\theta_1}\right)$$
$$\cdot p^Z[(0 \mid y_1, y_2, \ldots, y_J), (0 \mid x_1, x_2, \ldots, x_J)].$$

Inserting (3.23) into (3.24) yields for $x = (0 \mid x_1, x_2, \ldots, x_J)$

$$\left(\frac{q_1}{p_1}\right)^{x_1} \cdot \prod_{j=2}^{J} \left\{\left(\frac{1}{q_j}\right)^{\eta(0,x_j)} \left(\frac{q_j}{p_j}\right)^{x_j}\right\} \cdot q_1 \qquad (3.25)$$

$$= \sum_{\substack{y=(0|y_1,y_2,\ldots,y_J) \in S_1(C_0) \\ \exists B \subseteq A(y): H_B(y)=(0|x_1,x_2,\ldots,x_J)}} \left(\frac{q_1}{p_1}\right)^{y_1} \cdot \prod_{j=2}^{J} \left\{\left(\frac{1}{q_j}\right)^{\eta(0,y_j)} \cdot \left(\frac{q_j}{p_j}\right)^{y_j}\right\} \cdot q_1^{\delta(0,y_1)}$$

$$\cdot \left(\prod_{j \in B} p_j\right) \left(\prod_{j \in A(y)-B} q_j\right)$$

For $x_1 = 0$ the RHS is

$$q_1 \cdot \sum_{\substack{y=(0|0,y_2,\ldots,y_J) \in S_1(C_0) \\ \exists B \subseteq A(y): H_B(y)=x, J \notin B}} \prod_{j=2}^{J} \left\{\left(\frac{1}{q_j}\right)^{\eta(0,y_j)} \cdot \left(\frac{q_j}{p_j}\right)^{y_j}\right\}$$

$$\cdot \left(\prod_{j \in B} p_j\right) \cdot \left(\prod_{j \in A(y)-B} q_j\right)$$

$$+ \; q_1 \cdot \sum_{\substack{y=(0|1,y_2,\dots,y_J)\in S_1(C_0) \\ \exists B \subseteq A(y): H_B(y)=x, J \notin B, 1 \in B}} \frac{1}{q_1} \cdot \frac{q_1}{p_1} \cdot \prod_{j=2}^{J} \left\{ \left(\frac{1}{q_j}\right)^{\eta(0,y_j)} \cdot \left(\frac{q_j}{p_j}\right)^{y_j} \right\}$$

$$\cdot \left(\prod_{j\in B} p_j \right) \cdot \left(\prod_{j\in A(y)-B} q_j \right)$$

$$= \; q_1 \cdot \sum_{\substack{y=(0|y_1,y_2,\dots,y_J)\in S_1(C_0), \\ \exists B \subseteq A(y): H_B(y)=x}} \pi^{K-1,J}(y_1,y_2,\dots,y_J) \cdot G(K-1,J)$$

$$\cdot p[(y_1,y_2,\dots,y_J),(0,x_2,\dots,x_J)].$$

Cancelling the common factor q_1, on the RHS we find the *non-normalized equilibrium probability (in a cycle with J nodes and $K-1$ customers) that a 1-step transition leads to* $(0,x_2,\dots,x_J)$, which is just $\pi^{K-1,J}(0,x_2,\dots,x_J)$. For $x_1 > 0$ (noticing $y_1 > 0$!) (3.25) reads

$$q_1 \cdot \prod_{j=1}^{J} \left\{ \left(\frac{1}{q_j}\right)^{\eta(0,x_j)} \cdot \left(\frac{q_j}{p_j}\right)^{x_j} \right\} \cdot q_1 \tag{3.26}$$

$$= \sum_{\substack{y=(0|y_1,y_2,\dots,y_J)\in S_1(C_0) \\ \exists B \subseteq A(y): H_B(y)=x, J\notin B}} \prod_{j=1}^{J} \left\{ \left(\frac{1}{q_j}\right)^{\eta(0,y_j)} \cdot \left(\frac{q_j}{p_j}\right)^{y_j} \right\}$$

$$\cdot q_1 \cdot \left(\prod_{j\in B} p_j \right) \cdot \left(\prod_{j\in A(y)-B} q_j \right) \cdot$$

Cancelling in (3.26) the common factor q_1, we find on the RHS *the non-normalized steady-state probability (in a cycle with J nodes and $K-1$ customers) that a 1-step transition which does not show a departure from node J to node 1 leads to* $(x_1,x_2,\dots,x_J)$. Considering the time reversal of that system, we see that this is just the probability of a 1-step transition from $(x_1,x_2,\dots,x_J)$ without departure from node 1. This is the LHS of (3.26).

Our second step is inserting (3.22) into (3.21) for $\tilde{x}_1 = 1$. This yields for $x = (1 \mid x_1, x_2, \dots, x_J) \in S_1(C_0)$

$$\left(\frac{q_1}{p_1}\right)^{x_1+1} \cdot \prod_{j=2}^{J} \left\{ \left(\frac{1}{q_j}\right)^{\eta(0,x_j)} \cdot \left(\frac{q_j}{p_j}\right)^{x_j} \right\} \cdot \frac{p_1\theta}{1-q_1\theta} \cdot \frac{1}{1-q_1\theta} \tag{3.27}$$

$$= \; \theta \cdot \sum_{y=(\tilde{y}_1|y_1,y_2,\dots,y_J)\in S_1(C_0)} \left(\frac{q_1}{p_1}\right)^{\tilde{y}_1+y_1} \cdot \prod_{j=2}^{J} \left\{ \left(\frac{1}{q_j}\right)^{\eta(0,y_j)} \cdot \left(\frac{q_j}{p_j}\right)^{y_j} \right\}$$

$$\cdot \left(\frac{p_1\theta}{1-q_1\theta}\right)^{\tilde{y}_1} \cdot \frac{1}{1-q_1\theta} \cdot p^Z[(\tilde{y}_1 \mid y_1,y_2,\dots,y_J),(1 \mid x_1,x_2,\dots,x_J)].$$

For $x_1 = 0$ we have in (3.27)

$$\left(\frac{q_1}{p_1}\right) \cdot \prod_{j=2}^{J}\left\{\left(\frac{1}{q_j}\right)^{\eta(0,x_j)} \cdot \left(\frac{q_j}{p_j}\right)^{x_j}\right\} \cdot \frac{p_1\theta}{1-q_1\theta} \tag{3.28}$$

$$= \ \theta \cdot \sum_{\substack{y=(0|0,y_2,\dots,y_J)\in S_1(C_0) \\ \exists B\subseteq A(y):H_B(y)=x, J\in B}} \prod_{j=2}^{J}\left\{\left(\frac{1}{q_j}\right)^{\eta(0,y_j)} \cdot \left(\frac{q_j}{p_j}\right)^{y_j}\right\}$$

$$\cdot \left(\prod_{j\in B} p_j\right) \cdot \left(\prod_{j\in A(y)-B} q_j\right) \cdot q_1$$

$$+\theta \cdot \sum_{\substack{y=(0|1,y_2,\dots,y_J)\in S_1(C_0) \\ \exists B\subseteq A(y):H_B(y)=x,1,J\in B}} q_1 \cdot \left(\frac{1}{q_1}\right)\left(\frac{q_1}{p_1}\right)\cdot\prod_{j=2}^{J}\left\{\left(\frac{1}{q_j}\right)^{\eta(0,y_j)} \cdot \left(\frac{q_j}{p_j}\right)^{y_j}\right\}$$

$$\cdot \left(\prod_{j\in B} p_j\right) \cdot \left(\prod_{j\in A(y)-B} q_j\right)$$

$$+\theta \cdot \sum_{\substack{(1|0,y_2,\dots,y_J)\in S_1(C_0) \\ \exists B\subseteq A(y):H_B(y)=x,J\notin B}} \left(\frac{q_1}{p_1}\right)\cdot\prod_{j=2}^{J}\left\{\left(\frac{1}{q_j}\right)^{\eta(0,y_j)} \cdot \left(\frac{q_j}{p_j}\right)^{y_j}\right\} \cdot \frac{p_1\theta}{1-q_1\theta}$$

$$\cdot \left(\prod_{j\in B} p_j\right) \cdot \left(\prod_{j\in A(y)-B} q_j\right) \cdot q_1+$$

$$+\theta \cdot \sum_{\substack{(1|1,y_2,\dots,y_J)\in S_1(C_0) \\ \exists B\subseteq A(y):H_B(y)=x,1\in B,J\notin B}} q_1 \cdot \frac{1}{q_1} \cdot \left(\frac{q_1}{p_1}\right)^2 \cdot \prod_{j=2}^{J}\left\{\left(\frac{1}{q_j}\right)^{\eta(0,y_j)} \cdot \left(\frac{q_j}{p_j}\right)^{y_j}\right\}$$

$$\cdot \frac{p_1\theta}{1-q_1\theta} \cdot \left(\prod_{j\in B} p_j\right) \cdot \left(\prod_{j\in A(y)-B} q_j\right) \cdot$$

Now the first and the second summand of the RHS of (3.28) are $q_1\theta \cdot \frac{p_1}{q_1}$ times
the non–normalized equilibrium probability (in a J-node cycle with $K-1$ customers) that state $(1,x_2,\dots,x_J) \in S(K-1,J)$ is reached by at least a customer's
movement from node J to node 1. Considering the time reversal of that cycle we
obtain for this sum

$$q_1\theta \cdot \frac{1}{q_1} \cdot \frac{q_1}{p_1} \cdot \prod_{j=2}^{J}\left\{\left(\frac{1}{q_j}\right)^{\eta(0,x_j)}\left(\frac{q_j}{p_j}\right)^{x_j}\right\} \cdot p_1.$$

The third and the fourth summand of the RHS of (3.28) are $\theta^2 q_1^2/(1-q_1\theta)$
times the non–normalized equilibrium probability (in a system with J nodes and

$K-2$ customers) that state $(0, x_2, x_3, \ldots, x_J) \in S(K-2, J)$ is reached by 1-step transitions. From the equilibrium equation in that system this is

$$\frac{\theta^2 q_1^2}{1 - q_1\theta} \cdot \prod_{j=2}^{J} \left\{ \left(\frac{1}{q_j}\right)^{\eta(0,x_j)} \cdot \left(\frac{q_j}{p_j}\right)^{x_j} \right\}.$$

Summing up the RHS of (3.28) therefore yields

$$\frac{p_1\theta}{1 - q_1\theta} \cdot \frac{q_1}{p_1} \cdot \prod_{j=2}^{J} \left\{ \left(\frac{1}{q_j}\right)^{\eta(0,x_j)} \cdot \left(\frac{q_j}{p_j}\right)^{x_j} \right\},$$

which is the LHS of (3.28).
For $x_1 > 0$ we have in (3.27) (recall $\tilde{x}_1 = 1$):

$$\left(\frac{q_1}{p_1}\right)^{x_1+1} \cdot \prod_{j=2}^{J} \left\{ \left(\frac{1}{q_j}\right)^{\eta(0,x_j)} \cdot \left(\frac{q_j}{p_j}\right)^{x_j} \right\} \cdot \frac{p_1\theta}{1 - q_1\theta} \tag{3.29}$$

$$= \theta \cdot \sum_{\substack{y=(0|x_1,y_2,\ldots,y_J)\in S_1(C_0) \\ \exists B \subseteq A(y): H_B(y)=x, J\in B, 1\notin B}} \left(\frac{q_1}{p_1}\right)^{x_1} \cdot \prod_{j=2}^{J} \left\{ \left(\frac{1}{q_j}\right)^{\eta(0,y_j)} \cdot \left(\frac{q_j}{p_j}\right)^{y_j} \right\}$$

$$\cdot \left(\prod_{j\in B} p_j\right) \cdot \left(\prod_{j\in A(y)-B-\{1\}} q_j\right) \cdot q_1 +$$

$$+\theta \cdot \sum_{\substack{y=(0|x_1+1,y_2,\ldots,y_J)\in S_1(C_0) \\ \exists B \subseteq A(y): H_B(y)=x, 1, J\in B}} q_1 \cdot \left(\frac{1}{q_1}\right) \cdot \left(\frac{q_1}{p_1}\right)^{x_1+1}$$

$$\prod_{j=2}^{J} \left\{ \left(\frac{1}{q_j}\right)^{\eta(0,y_j)} \cdot \left(\frac{q_j}{p_j}\right)^{y_j} \right\} \cdot \left(\prod_{j\in B} p_j\right) \cdot \left(\prod_{j\in A(y)-B} q_j\right)$$

$$+\theta \cdot \sum_{\substack{y=(1|x_1,y_2,\ldots,y_J)\in S_1(C_0) \\ \exists B \subseteq A(y): H_B(y)=x, 1, J\notin B}} q_1 \cdot \left(\frac{1}{q_1}\right) \cdot \left(\frac{q_1}{p_1}\right)^{x_1+1}$$

$$\cdot \prod_{j=2}^{J} \left\{ \left(\frac{1}{q_j}\right)^{\eta(0,y_j)} \cdot \left(\frac{q_j}{p_j}\right)^{y_j} \right\} \cdot \frac{p_1\theta}{1 - q_1\theta} \cdot \left(\prod_{j\in B} p_j\right) \cdot \left(\prod_{j\in A(y)-B} q_j\right)$$

$$+\theta \cdot \sum_{\substack{y=(1|x_1+1,y_2,\ldots,y_J)\in S_1(C_0) \\ \exists B \subseteq A(y): H_B(y)=x, 1\in B, J\notin B}} q_1 \cdot \left(\frac{1}{q_1}\right) \cdot \left(\frac{q_1}{p_1}\right)^{x_1+2}$$

$$\cdot \prod_{j=2}^{J} \left\{ \left(\frac{1}{q_j}\right)^{\eta(0,y_j)} \cdot \left(\frac{q_j}{p_j}\right)^{y_j} \right\} \cdot \frac{p_1\theta}{1 - q_1\theta} \cdot \left(\prod_{j\in B} p_j\right) \cdot \left(\prod_{j\in A(y)-B} q_j\right).$$

Here the first and the second summand of the RHS is $q_1\theta \cdot \left(\frac{q_1}{p_1}\right)^{x_1}$ times *the non-normalized equilibrium probability (in a J-node cycle with $K - 1 - x_1$ customers cycling) that $(1, x_2, \ldots, x_J) \in S(K - 1 - x_1, J)$ is reached by (at least) a customer's movement from node J to node 1.* Considering the time reversal of that system we obtain for this sum

$$\left(\frac{q_1}{p_1}\right)^{x_1} \cdot \theta q_1 \cdot \frac{1}{q_1}\left(\frac{q_1}{p_1}\right) \cdot \prod_{j=2}^{J}\left\{\left(\frac{1}{q_j}\right)^{\eta(0,x_j))}\left(\frac{q_j}{p_j}\right)^{x_j}\right\} \cdot p_1.$$

The third and the fourth summand of the RHS of (3.29) are $\theta^2 q_1^2 / (1 - q_1\theta)$ times *the non–normalized equilibrium probability (in a J-node cycle with $K - 2$ customers cycling) that state $(x_1, x_2, \ldots, x_J) \in S(K - 2, J)$ is reached by transitions where departures from J are forbidden.* Considering the time reversal of that system this equals *the non-normalized equilibrium probability that state $(x_1, x_2, \ldots, x_J) \in S(K - 2, J)$ is left by transitions without the service at node 1 expiring,* i.e.,

$$\frac{\theta^2 q_1^2}{1 - q_1\theta} \cdot \frac{1}{q_1}\left(\frac{q_1}{p_1}\right)^{x_1} \cdot \prod_{j=2}^{J}\left\{\left(\frac{1}{q_j}\right)^{\eta(0,x_j)}\left(\frac{q_j}{p_j}\right)^{x_j}\right\} \cdot q_1.$$

Summing up the RHS of (3.29) yields the LHS.

Let us assume that we have proved (3.22) for all $f[y; \theta], y = (\tilde{y}_1 \mid y_1, y_2, \ldots, y_J)$ with $0 \le \tilde{y}_1 \le \tilde{x}_1 - 1, 2 \le \tilde{x}_1 < K - 1$. For induction we have to prove (3.22) for all $f[(\tilde{x}_1 \mid x_1, x_2, \ldots, x_J); \theta]$.

Inserting (3.22) into (3.21) this reads

$$\left(\frac{q_1}{p_1}\right)^{\tilde{x}_1+x_1} \cdot \prod_{j=2}^{J}\left\{\left(\frac{1}{q_j}\right)^{\eta(0,x_j)}\cdot\left(\frac{q_j}{p_j}\right)^{x_j}\right\} \cdot \left(\frac{p_1\theta}{1-q_1\theta}\right)^{\tilde{x}_1} \cdot \frac{1}{1-q_1\theta}$$

$$= \theta \cdot \sum_{(\tilde{y}_1 \mid y_1, y_2, \ldots, y_J) \in S_1(C_0)} \left(\frac{q_1}{p_1}\right)^{\tilde{y}_1+y_1} \cdot \prod_{j=2}^{J}\left\{\left(\frac{1}{q_j}\right)^{\eta(0,y_j)}\left(\frac{q_j}{p_j}\right)^{y_j}\right\} \quad (3.30)$$

$$\cdot \left(\frac{p_1\theta}{1-q_1\theta}\right)^{\tilde{y}_1} \cdot \frac{1}{1-q_1\theta} \cdot p^Z[(\tilde{y}_1 \mid y_1, y_2, \ldots, y_J), (\tilde{x}_1 \mid x_1, x_2, \ldots, x_J)].$$

For $x_1 = 0$, i.e., $x = (\tilde{x}_1 \mid 0, x_2, \ldots, x_J)$, this is

$$\left(\frac{q_1}{p_1}\right)^{\tilde{x}_1} \cdot \prod_{j=2}^{J}\left\{\left(\frac{1}{q_j}\right)^{\eta(0,x_j)}\left(\frac{q_j}{p_j}\right)^{x_j}\right\} \cdot \left(\frac{p_1\theta}{1-q_1\theta}\right)^{\tilde{x}_1} \quad (3.31)$$

$$= \theta \cdot \sum_{\substack{y=(\tilde{x}_1-1\mid0,y_2,\ldots,y_J)\in S_1(C_0) \\ \exists B \subseteq A(y): H_B(y)=x, J\in B}} \left(\frac{q_1}{p_1}\right)^{\tilde{x}_1-1} \cdot \prod_{j=2}^{J}\left\{\left(\frac{1}{q_j}\right)^{\eta(0,y_j)}\cdot\left(\frac{q_j}{p_j}\right)^{y_j}\right\}$$

$$\cdot \left(\frac{p_1\theta}{1-q_1\theta}\right)^{\tilde{x}_1-1} \cdot \left(\prod_{j\in B} p_j\right) \cdot \left(\prod_{j\in A(y)-B} q_j\right) \cdot q_1$$

$$+\theta \cdot \sum_{\substack{y=(\tilde{x}_1-1|1,y_2,\dots,y_J)\in S_1(C_0) \\ \exists B\subseteq A(y):H_B(y)=x,1,J\in B}} \left(\frac{q_1}{p_1}\right)^{\tilde{x}_1} \cdot \prod_{j=2}^{J}\left\{\left(\frac{1}{q_j}\right)^{\eta(0,y_j)} \cdot \left(\frac{q_j}{p_j}\right)^{y_j}\right\}$$

$$\cdot \left(\frac{p_1\theta}{1-q_1\theta}\right)^{\tilde{x}_1-1} \cdot \left(\prod_{j\in B} p_j\right) \cdot \left(\prod_{j\in A(y)-B} q_j\right)$$

$$+\theta \cdot \sum_{\substack{y=(\tilde{x}_1|0,y_2,\dots,y_J)\in S_1(C_0) \\ \exists B\subseteq A(y):H_B(y)=x,J\notin B}} \left(\frac{q_1}{p_1}\right)^{\tilde{x}_1} \cdot \prod_{j=2}^{J}\left\{\left(\frac{1}{q_j}\right)^{\eta(0,y_j)} \cdot \left(\frac{q_j}{p_j}\right)^{y_j}\right\}$$

$$\cdot \left(\frac{p_1\theta}{1-q_1\theta}\right)^{\tilde{x}_1} \cdot \left(\prod_{j\in B} p_j\right) \cdot \left(\prod_{j\in A(y)-B} q_j\right) \cdot q_1$$

$$+\theta \cdot \sum_{\substack{y=(\tilde{x}_1|1,y_2,\dots,y_J)\in S_1(C_0) \\ \exists B\subseteq A(y):H_B(y)=x,J\notin B,1\in B}} \left(\frac{q_1}{p_1}\right)^{\tilde{x}_1+1} \cdot \prod_{j=2}^{J}\left\{\left(\frac{1}{q_j}\right)^{\eta(0,y_j)} \cdot \left(\frac{q_j}{p_j}\right)^{y_j}\right\}$$

$$\cdot \left(\frac{p_1\theta}{1-q_1\theta}\right)^{\tilde{x}_1} \cdot \left(\prod_{j\in B} p_j\right) \cdot \left(\prod_{j\in A(y)-B} q_j\right) \cdot$$

Now the first and second summand of the RHS of (3.31) are $\theta q_1 \cdot \left(\frac{q_1}{p_1}\right)^{\tilde{x}_1-1} \cdot$ $\left(\frac{p_1\theta}{1-q_1\theta}\right)^{\tilde{x}_1-1}$ times *the non-normalized equilibrium probability that in a cycle of* J *nodes with* $K-\tilde{x}_1$ *customers cycling state* $(1,x_2,\dots,x_J) \in S(K-\tilde{x}_1,J)$ *is reached by (at least) a customer's movement from node* J *to node* 1.

Considering the time reversal of that cycle we obtain for this sum

$$\theta q_1 \cdot \left(\frac{q_1}{p_1}\right)^{\tilde{x}_1-1} \cdot \left(\frac{p_1\theta}{1-q_1\theta}\right)^{\tilde{x}_1-1} \cdot \frac{1}{q_1} \cdot \left(\frac{q_1}{p_1}\right) \cdot \prod_{j=2}^{J}\left\{\left(\frac{1}{q_j}\right)^{\eta(0,x_j)} \cdot \left(\frac{q_j}{p_j}\right)^{x_j}\right\} \cdot p_1.$$

The third and the fourth summand of the RHS of (3.31) are $\theta^2 q_1^2 \cdot \left(\frac{q_1}{p_1}\right)^{\tilde{x}_1-1} \cdot$ $\left(\frac{p_1\theta}{1-q_1\theta}\right)^{\tilde{x}_1-1} \cdot \frac{1}{1-q_1\theta}$ times *the non-normalized equilibrium probability (in a* J-*node cycle with* $K-\tilde{x}_1$ *customers cycling) that state* $(0,x_2,x_3,\dots,x_J) \in S(K-\tilde{x}_1,J)$ *is reached by transitions where departures from node* J *are forbidden.* Considering the time reversal of that system we obtain for this sum

$$\theta^2 q_1^2 \cdot \left(\frac{q_1}{p_1}\right)^{\tilde{x}_1-1} \cdot \left(\frac{p_1\theta}{1-q_1\theta}\right)^{\tilde{x}_1-1} \cdot \frac{1}{1-q_1\theta} \cdot \prod_{j=2}^{J}\left(\frac{1}{q_j}\right)^{\eta(0,x_j)} \left(\frac{q_j}{p_j}\right)^{x_j} \cdot$$

Summing up the RHS yields the LHS of (3.31).

For $x = (\tilde{x}_1 \mid x_1, x_2, \ldots, x_J)$ with $x_1 > 0$ we obtain

$$\left(\frac{q_1}{p_1}\right)^{\tilde{x}_1 + x_1} \cdot \prod_{j=2}^{J} \left\{ \left(\frac{1}{q_j}\right)^{\eta(0,x_j)} \cdot \left(\frac{q_j}{p_j}\right)^{x_j} \right\} \cdot \left(\frac{p_1 \theta}{1 - q_1 \theta}\right)^{\tilde{x}_1} \tag{3.32}$$

$$= \theta \cdot \sum_{\substack{y = (\tilde{x}_1 - 1 \mid x_1, y_2, \ldots, y_J) \in S_1(C_0) \\ \exists B \subseteq A(y): H_B(y) = x, J \in B, 1 \notin B}} \left(\frac{q_1}{p_1}\right)^{\tilde{x}_1 - 1 + x_1} \cdot \prod_{j=2}^{J} \left\{ \left(\frac{1}{q_j}\right)^{\eta(0,x_j)} \left(\frac{q_j}{p_j}\right)^{x_j} \right\}$$

$$\cdot \left(\frac{p_1 \theta}{1 - q_1 \theta}\right)^{\tilde{x}_1 - 1} \cdot \left(\prod_{j \in B} p_j\right) \cdot \left(\prod_{j \in A(y) - B - \{1\}} q_j\right) \cdot q_1$$

$$+ \; \theta \cdot \sum_{\substack{y = (\tilde{x}_1 - 1 \mid x_1 + 1, y_2, \ldots, y_J) \in S_1(C_0) \\ \exists B \subseteq A(y): H_B(y) = x, 1, J \in B}} \left(\frac{q_1}{p_1}\right)^{\tilde{x}_1 + x_1} \cdot \prod_{j=2}^{J} \left\{ \left(\frac{1}{q_j}\right)^{\eta(0,x_j)} \cdot \left(\frac{q_j}{p_j}\right)^{x_j} \right\}$$

$$\cdot \left(\frac{p_1 \theta}{1 - q_1 \theta}\right)^{\tilde{x}_1 - 1} \cdot \left(\prod_{j \in B} p_j\right) \cdot \left(\prod_{j \in A(y) - B} q_j\right)$$

$$+ \; \theta \cdot \sum_{\substack{y = (\tilde{x}_1 \mid x_1, y_2, \ldots, y_J) \in S_1(C_0) \\ \exists B \subseteq A(y): H_B(y) = x, 1, J \notin B}} \left(\frac{q_1}{p_1}\right)^{\tilde{x}_1 + x_1} \cdot \prod_{j=2}^{J} \left\{ \left(\frac{1}{q_j}\right)^{\eta(0,x_j)} \cdot \left(\frac{q_j}{p_j}\right)^{x_j} \right\}$$

$$\cdot \left(\frac{p_1 \theta}{1 - q_1 \theta}\right)^{\tilde{x}_1} \cdot \left(\prod_{j \in B} p_j\right) \cdot \left(\prod_{j \in A(y) - B} q_j\right) +$$

$$+ \; \theta \cdot \sum_{\substack{y = (\tilde{x}_1 \mid x_1 + 1, y_2, \ldots, y_J) \in S_1(C_0) \\ \exists B \subseteq A(y): H_B(y) = x, 1 \in B, J \notin B}} \left(\frac{q_1}{p_1}\right)^{\tilde{x}_1 + x_1 + 1} \cdot \prod_{j=2}^{J} \left\{ \left(\frac{1}{q_j}\right)^{\eta(0,x_j)} \cdot \left(\frac{q_j}{p_j}\right)^{x_j} \right\}$$

$$\cdot \left(\frac{p_1 \theta}{1 - q_1 \theta}\right)^{\tilde{x}_1} \cdot \left(\prod_{j \in B} p_j\right) \cdot \left(\prod_{j \in A(y) - B} q_j\right).$$

Here the first and the second summand of (3.32) are $q_1 \theta \cdot \left(\frac{q_1}{p_1}\right)^{\tilde{x}_1 - 1 + x_1} \cdot \left(\frac{p_1 \theta}{1 - q_1 \theta}\right)^{\tilde{x}_1 - 1}$ times *the non-normalized equilibrium probability (in a cycle of J nodes with $K - \tilde{x}_1 - x_1$ customers cycling) that $(1, x_2, \ldots, x_J) \in S(K - \tilde{x}_1 - x_1, J)$ is reached by (at least) a customer's movement from node J to node 1.* Considering the time reversal of that system we obtain for this sum

$$q_1 \theta \cdot \left(\frac{q_1}{p_1}\right)^{\tilde{x}_1 - 1 + x_1} \cdot \left(\frac{p_1 \theta}{1 - q_1 \theta}\right)^{\tilde{x}_1 - 1} \cdot \left(\frac{1}{q_1} \cdot \frac{q_1}{p_1}\right) \prod_{j=2}^{J} \left\{ \left(\frac{1}{q_j}\right)^{\eta(0,x_j)} \left(\frac{q_j}{p_j}\right)^{x_j} \right\} \cdot p_1.$$

The third and the fourth summand of (3.32) are $\theta q_1 \cdot \left(\frac{q_1}{p_1}\right)^{\tilde{x}_1} \cdot \left(\frac{p_1 \theta}{1 - q_1 \theta}\right)^{\tilde{x}_1}$ times *the non-normalized equilibrium probability (in a J-node cycle with $K - \tilde{x}_1 - 1$*

*customers cycling) that state $(x_1, x_2, \ldots, x_J) \in S(K - \tilde{x}_1 - 1, J)$ is reached by
1-step transitions with customer movements from J to 1 forbidden. Considering
the time reversal of that system we obtain for this sum*

$$\theta q_1 \cdot \left(\frac{q_1}{p_1}\right)^{\tilde{x}_1} \cdot \left(\frac{p_1 \theta}{1 - q_1 \theta}\right)^{\tilde{x}_1} \cdot \prod_{j=1}^{J} \left\{ \left(\frac{1}{q_j}\right)^{\eta(0, x_j)} \cdot \left(\frac{q_j}{p_j}\right)^{x_j} \right\} \cdot q_1.$$

Collecting both sums yields the LHS of (3.32).

We therefore have proved (3.22). Applying now (3.20) yields

$$\sum_{t=1}^{\infty} P(S_1 = t, Z(S_1) = (x_1, 0 \mid x_2, x_3, \ldots, x_J)) \cdot \theta^t \tag{3.33}$$

$$= \theta \cdot \sum_{t=0}^{\infty} \theta^t \cdot \sum_{(\tilde{y}_1 \mid 0, y_2, y_3, \ldots, y_J) \in S_1(C_0)} P(Z(t) = (\tilde{y}_1 \mid 0, y_2, y_3, \ldots, y_J))$$

$$\cdot p^Z[(\tilde{y}_1 \mid 0, y_2, y_3, \ldots, y_J), (x_1, 0 \mid x_2, x_3, \ldots, x_J)]$$

$$= \theta \cdot \sum_{\substack{y = (\tilde{y}_1 \mid 0, y_2, y_3, \ldots, y_J) \in S_1(C_0), \exists B \subseteq A(y): \\ x_1 = \tilde{y}_1 + 1\{J \in B\}, x_j = y_j + 1\{j - 1 \in B\} - 1\{j \in B\}, j = 2, \ldots, J}} \pi_1^{K, J}(\tilde{y}_1, y_2, y_3, \ldots, y_J)$$

$$\cdot \left(\frac{p_1 \theta}{1 - q_1 \theta}\right)^{\tilde{y}_1} \cdot \frac{1}{1 - q_1 \theta} \cdot \left(\prod_{j \in B} p_j\right) \cdot \left(\prod_{j \in A(y) - B} q_j\right) \cdot p_1.$$

(Note that from the definition $1 \notin A(y)$. $1\{\cdot\}$ denotes indicator functions.)

For $x_1 = 0$ the RHS of (3.33) is

$$\frac{p_1 \theta}{1 - q_1 \theta} \cdot \sum_{\substack{y \in (0 \mid 0, y_2, \ldots, y_J) \in S_1(C_0), \exists B \subseteq A(y): J \notin B, \\ x_j = y_j + 1\{j - 1 \in B\} - 1\{j \in B\}, j = 2, \ldots, J}} \prod_{j=2}^{J} \left\{ \left(\frac{1}{q_j}\right)^{\eta(0, y_j)} \cdot \left(\frac{q_j}{p_j}\right)^{y_j} \right\}$$

$$\cdot \left(\prod_{j \in B} p_j\right) \cdot \left(\prod_{j \in A(y) - B} q_j\right). \tag{3.34}$$

This is $p_1 \theta / (1 - q_1 \theta)$ times *the non-normalized equilibrium probability (in a cycle
of nodes $2, 3, \ldots, J$ with $K - 1$ customers cycling) that state $(x_2, x_3, \ldots, x_J)$ is
reached with departures from node J forbidden.* Considering the time reversal of
that system, we see that this is *the non-normalized probability of a 1-step transition from $(x_2, x_3, \ldots, x_J)$ with departures from 2 to J forbidden.* This yields

$$(3.34) = \frac{p_1 \theta}{1 - q_1 \theta} \cdot \prod_{j=2}^{J} \left\{ \left(\frac{1}{q_j}\right)^{\eta(0, x_j)} \cdot \left(\frac{q_j}{p_j}\right)^{x_j} \right\} \cdot q_2^{\eta(0, x_2)},$$

which is for $x_1 = 0$ the RHS of (3.15).

For $x_1 > 0$ the RHS of (3.33) is

$$
\left(\frac{p_1\theta}{1-q_1\theta}\right)^{x_1} \cdot \left[\sum_{\substack{y=(x_1-1|0,y_2,\ldots,y_J)\in S_1(C_0),\exists B\subseteq A(y):J\in B,\\ x_j=v_j+1\{j-1\in B\}-1\{j\in B\},j=2,\ldots,J}} \left(\frac{q_1}{p_1}\right)^{x_1-1} \right. \tag{3.35}
$$

$$
\prod_{j=2}^{J}\left\{\left(\frac{1}{q_j}\right)^{\eta(0,y_j)}\cdot\left(\frac{q_j}{p_j}\right)^{y_j}\right\}\cdot\left(\prod_{j\in B}p_j\right)\cdot\left(\prod_{j\in A(y)-B}q_j\right)
$$

$$
+ \sum_{\substack{y=(x_1|0,y_2,\ldots,y_J)\in S_1(C_0),\exists B\subseteq A(y):J\notin B,\\ x_j=v_j+1\{j-1\in B\}-1\{j\in B\},j=2,\ldots,J}} \left(\frac{q_1}{p_1}\right)^{x_1}\cdot\prod_{j=2}^{J}\left\{\left(\frac{1}{q_j}\right)^{\eta(0,y_j)}\cdot\left(\frac{q_j}{p_j}\right)^{y_j}\right\}
$$

$$
\left.\cdot\left(\frac{p_1\theta}{1-q_1\theta}\right)\cdot\left(\prod_{j\in B}p_j\right)\cdot\left(\prod_{j\in A(y)-B}q_j\right)\right].
$$

Cancelling $\left(\frac{1}{q_J}\right)^{\eta(0,x_J)}\cdot\left(\frac{q_J}{p_J}\right)^{x_J}\cdot p_J = \left(\frac{1}{q_J}\right)^{\eta(0,x_J-1)}\cdot\left(\frac{q_J}{p_J}\right)^{x_J-1}\cdot q_J^{\eta(0,x_J-1)}$ in the first sum in the squared brackets of (3.35) we see that this is $\left(\frac{q_1}{p_1}\right)^{x_1-1}$ times *the non-normalized equilibrium probability (in a cycle of nodes* $2,3,\ldots,J$ *with* $K-1-x_1$ *customers cycling) that state* $(x_2,x_3,\ldots,x_J)\in S(K-1-x_1,J-1)$ *is reached by a 1-step transition where departures from* J *to* 2 *are forbidden.* Considering the time reversal of that system this turns out to be

$$
\left(\frac{q_1}{p_1}\right)^{x_1-1}\cdot\prod_{j=2}^{J}\left\{\left(\frac{1}{q_j}\right)^{\eta(0,x_j)}\cdot\left(\frac{q_j}{p_j}\right)^{x_j}\right\}\cdot q_2^{\eta(0,x_2)}.
$$

The second sum in the squared bracket of (3.35) is $\left(\frac{q_1}{p_1}\right)^{x_1}\cdot\frac{p_1\theta}{1-q_1\theta}$ times *the non-normalized equilibrium probability (in a cycle of nodes* $2,3,\ldots,J$ *with* $K-1-x_1$ *customers cycling) that state* $(x_2,x_3\ldots,x_J)\in S(K-1-x_1,J-1)$ *is reached by 1-step transitions where departures from* J *to* 2 *are forbidden.* Considering the time reversal of that system this turns out to be

$$
\left(\frac{q_1}{p_1}\right)^{x_1}\cdot\frac{p_1\theta}{1-q_1\theta}\cdot\prod_{j=2}^{J}\left\{\left(\frac{1}{q_j}\right)^{\eta(0,x_j)}\cdot\left(\frac{q_j}{p_j}\right)^{x_j}\right\}\cdot q_2.
$$

Collecting now all the terms of (3.35) yields for the case $x_1 > 0$ the RHS of (3.15).

We have proved Lemma 3.13.   ⊙

3.4 Computational Algorithms for Closed Cycles of State Independent Bernoulli Servers

It is well known that in performance analysis of continuous time systems to compute norming constants similar to those appearing in theorem 3.2 poses considerable difficulties. Several algorithms are developed for efficiently evaluating norming constants, steady state probabilities, and further quantities derived from this, which describe the quality of service provided by networks. In [13] classical results on algorithms to compute performance indices for closed networks based on norming constants and based on Mean Value Analysis (MVA) are described. A detailed analysis is presented in [5] of convolution algorithm (norming constants), MVA, and the more recent RECAL method, and additionally of approximation algorithms for product form networks as well as for non product form networks based on the concepts provided by these prototype algorithms. For a recent discussion and a short survey, including especially approximation procedures(which can be developed in the framework dealt with here in parallel), see [53]
Special algorithms for discrete time systems with finite capacity, which are with respect to some properties equivalent to closed networks, are developed in [125].

The structure of these algorithms point out the way to compute the interesting quantities here as well. We sketch in this section how to do this. A first observation is based on theorem 3.2 and proposition 3.5: While in continuous time models the norming constants in the steady state and in the arrival probabilities are of the same structure, we have to apply different computational schemes here. Clearly this holds for the multichain case (section 3.2) as well. For a concise presentation we restrict ourself to the unichain case, and assume the service probabilities to be state independent.

Proposition 3.17 (Norming constants) *(a) For the steady state distribution (3.2) in theorem 3.2 (with state independent service probabilities) the norming constant in a system with K customers cycling in J nodes is*

$$G(K,J) = \sum_{\substack{(n_1,\ldots,n_J)\in I\!\!N^J \\ n_1+\cdots+n_J=K}} \prod_{j=1}^{J} \left(\frac{q_j}{p_j}\right)^{n_j} \left(\frac{1}{q_j}\right)^{\eta(0,n_j)} , \quad K \geq 1, J \geq 1.$$

(b) The norming constant for the arrival probabilities (3.8) in proposition 3.5 seen by a type m–customer on his arrival at node i in a system with K customers cycling in J nodes is

$$G_{i,m}(K,J) = \sum_{\substack{(n_1,\ldots,n_J)\in I\!\!N^J \\ n_1+\cdots+n_J=K-1}} \left(\frac{q_i}{p_i}\right)^{n_i} \prod_{\substack{j=1 \\ j\neq i}}^{J} \left(\frac{q_j}{p_j}\right)^{n_j} \left(\frac{1}{q_j}\right)^{\eta(0,n_j)} , \quad K \geq 1, J \geq 1.$$

These norming constants are type independent. We set

$$G_i(K, J) := G_{i,m}(K, J), \quad m \in M.$$

The following is an analogue of *Buzen's algorithm* ([17]).

Proposition 3.18 (Buzen's algorithm) *For $K \geq 1, J \geq 1$ and $p_j \in (0,1)$,* $q_j = 1 - p_j, j = 1, \ldots, J$ *let*

$$G(K, J) = \sum_{\substack{(n_1, \ldots, n_J) \in \mathbb{N}^J \\ n_1 + \cdots + n_J = K}} \prod_{j=1}^{J} \left(\frac{q_j}{p_j}\right)^{n_j} \left(\frac{1}{q_j}\right)^{\eta(0, n_j)}.$$

Then the following recursion holds:

$$G(1, J) = \sum_{j=1}^{J} \frac{1}{p_j}, \qquad J \geq 1,$$

$$G(K, 1) = \left(\frac{q_1}{p_1}\right)^K \frac{1}{q_1}, \qquad K \geq 1,$$

$$G(K, J) = G(K, J - 1) + \frac{q_J}{p_J} G(K - 1, J) + G(K - 1, J - 1),$$
$$K \geq 2, J \geq 2. \tag{3.36}$$

Note that the computation of (3.36) depends on the prescribed numbering of the nodes. Renumbering the nodes leads to different paths for the algorithm.

The norming constant for the arrival probabilities will be computed for any customer arriving at node 1. It does not depend on the type of the arrival. Formally for arrivals at other nodes an algorithm similar to the following applies. But to make this distinction is not necessary: From lemma 7.3 it follows that for all $j = 1, \ldots, J$ equality $G_i(K, J) = G_1(K, J)$ holds.

Proposition 3.19 *For $K \geq 1, J \geq 1$ and $p_j \in (0, 1), q_j = 1 - p_j, j = 1, \ldots, J$* *let*

$$G_1(K, J) = \sum_{\substack{(n_1, \ldots, n_J) \in \mathbb{N}^J \\ n_1 + \cdots + n_J = K - 1}} \left(\frac{q_1}{p_1}\right)^{n_1} \prod_{j=2}^{J} \left(\frac{q_j}{p_j}\right)^{n_j} \left(\frac{1}{q_j}\right)^{\eta(0, n_j)}.$$

Then the following recursion holds:

$$G_1(2, J) = \frac{q_1}{p_1} + \sum_{j=2}^{J} \frac{1}{p_j}, \qquad J \geq 1,$$

$$G_1(K, J) = G(K - 1, J - 1) + \frac{q_J}{p_J} G_1(K - 1, J), \qquad K \geq 3, J \geq 1.$$

For systems in continuous time with state dependent service rates computational algorithms are described in [13]. Similar constructions can be proved in the discrete time setting along the lines sketched here. Some elementary consequences of the above algorithms follow.

Corollary 3.20 *For the closed cycle with state independent service probabilities in equilibrium $\pi^{K,J}$ (see theorem 3.7) we have:*
(a) The probability for the queue length X_1 at node 1 to exceed $k \in \{0, 1, \ldots, K\}$ is

$$P(X_1 \geq k) = \left(\frac{q_1}{p_1}\right)^k \frac{1}{q_1} G_1(K - k, J) G(K, J)^{-1}$$

(b) The mean steady state queue length $E[X_1]$ at node 1 is

$$E[X_1] = \frac{1}{q_1} G(K, J)^{-1} \sum_{k=1}^{K} \left(\frac{q_1}{p_1}\right)^k G_1(K - k, J).$$

(c) For $j = 1, \ldots, J$ and $k \geq 0$ we have

$$P(X_j \geq k) = P(X_1 \geq k) \left(\frac{p_1 q_j}{q_1 p_j}\right)^k \frac{q_1}{q_j}$$

Note that computing these performance indices needs an interplay of constants of different type: from time stationary steady states and from customer stationary steady states. Such distinction is not necessary in the continuous time setting.

Chapter 4

Open Tandems of State Dependent Bernoulli Servers with Different Customer Types

In this chapter we investigate a subclass of open networks of Bernoulli servers with simple topology: Series of linear ordered nodes, fed by an external arrival stream, i.e. open tandems. These are the analogues of tandems of exponential single server queues in continuous time under FCFS–regime. These tandem networks are of importance for many applications: Production lines, transmission lines in a telecommunication network, etc.

The classical results in the present context on steady state behaviour for such systems (with state independent Bernoulli input and indistinguishable customers) date back to 1976 – see [65]. Hsu and Burke solved the steady state problem by proving time reversibility of the nodes' local state process in equilibrium and then using induction. A similar procedure is not possible in the more general case considered here, because we have state dependent arrival processes which implies that departure streams are not homogeneous.

We describe the steady state behaviour of state dependent tandems with different customer types in section 4.2 and derive mean value performance indices. In section 4.3 we compute end–to–end–delay distribution for a customer traversing the tandem and the distribution of this customer's joint sojourn times at the different nodes of his itinerary.

Before presenting these results we describe (following [28]) in section 4.1 how these results can be applied to networks with more general topoloy, modeling e.g. transmission lines in meshed networks. This widens considerably the applicability of the seemingly narrow class of linear tandem systems.

H. Daduna: Queueing Networks with Discrete Time Scale, LNCS 2046, pp. 53–72, 2001.

In section 4.4 we consider an important related model: Re–entrant lines which
are the models for manufacturing systems with intermediate production control
and rework stages. So possibly parts have to reenter the same machine more
than once. According to the observed quality of the production an item on leave
from a production stage is either released to the next stage (if the quality is
sufficient high) or immediately sent back to the previous stage for rework. The
nodes which model the stages of the production process are the state dependent
feedback Bernoulli servers of section 2.3. We consider both the systems with and
without capacity constraints, which leads to closed cycles or open tandems of
feedback queues.

4.1   The Principle of Adjusted Transfer Rates

The aim of this section is to develop a simplified model for performance evaluation of a virtual channel in a meshed network, i.e., a transmission line prescribed
for a specific connection (see figure 1).

   The analoguous method for modeling linear substructures in general networks
is well established in the framework of continuous time systems: The principle of
adjusted transfer rates. This principle turns out to be of even more value in the
discrete time framework because nonlinear networks in general do not exhibit a
product form steady state and the equilibrium distributions are up to now not
available [56].

   Let us consider a virtual channel in a meshed communication network (figure
4.1). A virtual channel is a line of successive transmission channels dedicated to
a sender–receiver pair of nodes for a communication session. Although not necessarily a shortest connection such a virtual channel can be seen as a linear
substructure of the network. The nodes of this prescribed line of interest repre-
sent the transmission channels; they are numbered 1 through J. Calls of different
types arrive at node 1 according to a state dependent random process and pass
the nodes in increasing order. (The details for the dependence of the arrivals on
the network's state will be described below.)

   The calls are divided into cells of fixed (equal) length. The cells are transmitted according to a First–Come–First–Served regime one by one through the
successive transmission channels. The time for transmitting one cell through a
channel constitutes the inherent discrete time scale of the system: Transmission
of a cell needs exactly one time unit. Because it is assumed that the sender at
node 1 of the virtual channel transmits at most one cell per time unit into the
transmission line of nodes $1, \ldots, J$ no queues would built up if the line is used
only by such cells arriving at node 1, which are henceforth called regular cells.
Delay and queueing are due to the traffic from the rest of the network which
partly interferes with the regular cells at the nodes of the line. We incorporate
the effect of this background traffic into the model as follows:

   In a first step all background traffic originating from the rest of the network
that arives at node j is bundled (figure 4.2) into one stream of background cells

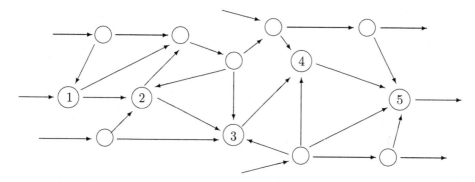

Fig. 4.1. Virtual channel with $J = 5$ nodes in a meshed network

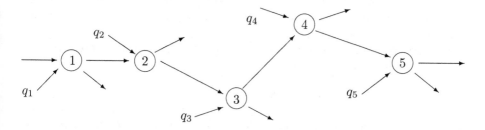

Fig. 4.2. The virtual channel with bundled background traffic

for node j, which interferes here with the regular cells from the virtual channel and thereafter leaves the line directly, $j = 1, \ldots, J$.

Therefore a regular cell on its arrival at node j will possibly find there a queue of other regular cells and of cells from the background traffic. The length of this queue is random due to the randomness of the arrival process of regular cells at node 1 and of the background traffic. From the point of view of regular cells they have to share the total capacity of the nodes (one cell per time unit) with cells of the background traffic. If the intensity ($=$ *mean number of arrivals*) of the background traffic at node j is $q_j \in [0, 1)$, then the capacity of node j dedicated to regular cells is $p_j := 1 - q_j$ cells per time unit.

In a second step the load of the background traffic on node j is locally incorporated into the model by reducing the capacity of node j by an amount of q_j to p_j cells per time unit (adjusted transfer rate), and thereafter neglecting the rest of the network outside of the virtual channel. This yields a linear network of nodes $1, \ldots, J$ with regular cells only (see figure 4.3).

Our resulting model (see figure 4.4) is therefore a transmission line of nodes 1 through J, where the capacity of node j is reduced from one cell per time unit to p_j cells per time unit, $0 < p_j \leq 1$. This is equivalent to having a mean transition time of $1/p_j \geq 1$ at node j. The randomness of the background traffic is incorporated by assuming geometrically distributed transition times with exactly

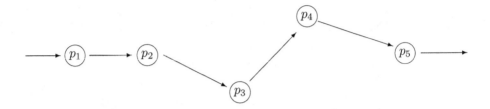

Fig. 4.3. The virtual channel as a line with reduced capacities

these means. It follows: A (regular) cell in transmission at the beginning of a time slot at station j will be completely transmitted at the end of this slot with probability p_j, with probability $1 - p_j$ the transmission will need at least one more time unit. Our aim is to develop an analytical framework for the following optimization problem:

Suppose the traffic of interest arrives as a Bernoulli stream of intensity $B \in (0,1]$ at station 1. Then find a control

$$\beta : I\!N \longrightarrow [0,1], \qquad n \longrightarrow \beta(n),$$

which constitutes a state dependent Bernoulli switch at the entrance point of the line. To be more specific: If the total load of the line is n (= sum of all queue lengths in the line), an arriving cell will be admitted with probability $\beta(n)$, with probability $1 - \beta(n)$ it will be rejected and will be lost. Consequently the arrival probability is $b(n) = B \cdot \beta(n)$. The control is to be chosen such that

(a) For a prescribed (small) probability $\gamma \in (0,1)$ the loss probability of a cell is less than γ;

(b) For a prescribed (large) probability $\alpha \in (0,1)$ and a critical transmission time t_0 the probability that transmission of a cell is performed within time t_0 is greater than α.

(c) The throughput is maximal given constraints (a) and (b).

This admission control scheme is incorporated in a general model which additionally allows to control locally the transmission capacity adapted to the local load. So in our model we have the following control options:

(i) The probability of terminating at node j a cell's transmission at the end of the present time unit may depend on the actual load $n_j > 0$ of that node (*queue length at node* j), $p_j = p_j(n_j), j = 1, \ldots, J$, and

(ii) the probability $b \in [0,1]$ of a new arrival at node 1 at the end of the present time unit may depend on the total load n of the system ($n = n_1 + \cdots + n_J$, the *total population size of the line*), $b = b(n)$, and on the type of the customer as will be specified below.

These control options for discrete time networks were investigated in the case of indistinguishable customers (cells) in [31] and include as an extreme case the classical window flow control, see [110], or loss systems. Introducing different

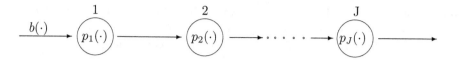

Fig. 4.4. Transmission line with adjusted state dependent capacities

customer types (cell types, [28]) turns out to be technically more involved but offers a more versatile modeling feature, e.g. for keeping trace of individual cells' behaviour and computing type dependent loss probabilities.

The *principle of adjusted transfer rates* in the continuous time framework is theoretically investigated by Reise, [110], [111], in connection with closed product form networks and especially their Mean Value Analysis (MVA). He proved that the steady state analysis in the reduced system yields the correct results even for the general network model. Results for open and mixed networks, especially concerning the analysis of delay time distributions can be found in [34]. Applying the principle of adjusted transfer rates to these models therefore provides exact analytical performance measures in the context of continuous time product form networks. In the discrete time setting of single server FCFS nodes a similar result allowing general network topology seems to be impossible, because nonlinear networks in general exhibit no product form equilibrium. But nevertheless it seems to be a reasonable approximation yielding explicit performance quantities and explicit measures for the quality of service, which are easily accessible.

4.2 Steady State and Arrival Theorem

We use Bernoulli servers as described in section 2.2 as building blocks of an open tandem of queues. The node characteristics remain the same as in the previous chapters. Customers of different types arrive according to a state dependent Bernoulli process at node 1, are served there according to FCFS, and proceed then through the sequence of nodes possibly changing their types, which is determined by the regime described in section 2.2. After leaving node J, the customers depart from the system.

We assume that there is only one chain of customers, all customers share the same type set M (see section 3.1 and 3.2 for a discussion). The entrance type of an arriving customer is chosen according to rules to be given below. If a customer of type $m \in M$ leaves node j, then this customer's type is resampled according to the probability matrix $r(j)$, his new type is $m' \in M$ with probability $r(j; m, m')$, $j = 1, \ldots, J$.

Regulation of simultaneous events is according to *late arrivals* and *departure before arrivals*, see [51].

The external arrival probabilities depend on the total population of the system and on the type of the arrival, i.e., if at time $t \in I\!N$ there are n_j customers present at node j, $j = 1, \ldots, J$, then a new arrival of type m appears in $(t, t+1]$ with probability $b(n_1 + \cdots + n_J) \cdot a(m) \in (0,1)$.

Service times and arrival decisions are conditionally independent given the actual vector of customer types at the nodes.

We use the definitions of section 2.2 and section 3.1: $M(i)$ denotes the set of possible types which customers may show when staying in node $i, i = 1, \ldots, J$. Here $M(1) = \{m \in M : a(m) > 0\}$, while $M(i), i = 2, \ldots, J$, is determined by solving the equation (3.1) for $\eta(i; \cdot), i = 2, \ldots, J$, in the present context and then setting $M(i) = \{m \in M : \eta(i; m) > 0\}$.

A typical state of the system is $x = (x_1, \ldots, x_J)$, where $x_j = e_j$ or $x_j = (x_{j1}, \ldots, x_{jn_j}) \in M(j)^{n_j}, 1 \le n_j, j = 1, \ldots, J$. x_j is called a local state for node j with the interpretation given before in section 3.1. These local states allow to construct the state space

$$\tilde{S}(J) := \{(x_1, \ldots, x_J) : x_j = (x_{j1}, \ldots, x_{jn_j}) \in M(j)^{n_j}, n_j \ge 0, j = 1, \ldots, J\},$$

where the process X, which desribes the time evolution of the network, is living on. Let $X_j(t)$ denote the local state at node j at time t, and $X(t) = (X_1(t), \ldots, X_J(t))$ the joint *vector of type sequences* at time t. $X = (X(t) : t \in I\!N)$ is a discrete time Markov chain with state space $\tilde{S}(J)$ and transition matrix $p = (p(x, y) : x, y \in \tilde{S}(J))$. X is irreducible on $\tilde{S}(J)$.

We prove stabilisation conditions for this tandem system starting from a closed cycle which comprises the given nodes and one extra node, which simulates the source and the sink, and then let the population size grow to infinity [31].

Theorem 4.1 (Steady state) *The Markov chain X is ergodic if and only if*

$$H(J) = \sum_{(n_1, \ldots, n_J) \in I\!N^J} \left(\frac{\prod_{h=0}^{n_1 + \cdots + n_J - 1} b(h)}{\prod_{h=0}^{n_1 + \cdots + n_J} c(h)} \right) \prod_{j=1}^{J} \left(\frac{\prod_{h=1}^{n_j - 1} q_j(h)}{\prod_{h=1}^{n_j} p_j(h)} \right) < \infty. \quad (4.1)$$

If this ergodicity condition is fulfilled, then the unique equilibrium distribution of X is $\pi^J = (\pi^J(x) : x \in \tilde{S}(J))$ given by

$$\pi^J(x_1, \ldots, x_J) \quad (4.2)$$
$$= \pi^J((x_{11}, \ldots, x_{1n_1}); \ldots; (x_{J1}, \ldots, x_{Jn_J})) =$$

$$= \left(\frac{\prod_{h=0}^{n_1 + \cdots + n_J - 1} b(h)}{\prod_{h=0}^{n_1 + \cdots + n_J} c(h)} \right) \prod_{j=1}^{J} \left(\prod_{k=1}^{n_j} \eta(j; x_{jk}) \right) \left(\frac{\prod_{h=1}^{n_j - 1} q_j(h)}{\prod_{h=1}^{n_j} p_j(h)} \right) \cdot H(J)^{-1},$$
$$(x_1, \ldots, x_J) \in \tilde{S}(J),$$

where $c(h) := 1 - b(h), h \in I\!N$.

Proof : Consider a closed cycle with K customers with common type set M, and nodes $0, 1, 2, \ldots, J$, where nodes $1, 2, \ldots, J$ are as described in section 3.1 with state dependent service probabilities $p_j(n_j)$ and and type selection matrices $r(j)$, $1 \leq j \leq J$.

Node 0 is an extra node simulating the traffic source: Customers present at node 0 are all of the same type, say m_0 and $r(J; m, m_0) = 1, m \in M$. We set $r(0; m_0, m) = a(m) = \eta(1; m)$ for $m \in M(1)$, and $r(0; m', \cdot)$ to be any probability on M for $m' \neq m_0$.

The service probability at node 0 is $p_0(n_0) = b(K - n_0), 1 \leq n_0 \leq K$. So, if this node 0 is occupied by the total population of size K, then the transmission line $1, \ldots, J$ is empty and node 1 observes a Bernoulli arrival stream with probability $b(0)$ for an arrival per time slot. An arriving customer at node 1 will be of type $m \in M$ with probability $a(m)$.

The steady–state distribution of this system (with $G(K, J+1)^{-1}$ the norming constant and n_0 the number of customers present at node 0) is by theorem 3.2

$$
\pi^{K,J+1}(n_0, (x_{11}, \ldots, x_{1n_1}); \ldots; (x_{J1}, \ldots, x_{Jn_J}))
$$

$$
= \left(\frac{\prod_{h=1}^{n_0-1} c(K-h)}{\prod_{h=1}^{n_0} b(K-h)} \right) \prod_{j=1}^{J} \left(\prod_{k=1}^{n_j} \eta(j; x_{jk}) \right) \left(\frac{\prod_{h=1}^{n_j-1} q_j(h)}{\prod_{h=1}^{n_j} p_j(h)} \right) \cdot G(K, J+1)^{-1}
$$

$$
= \left(\frac{\prod_{h=0}^{n_1+\cdots+n_J-1} b(h)}{\prod_{h=0}^{n_1+\cdots+n_J} c(h)} \right) \left(\prod_{h=0}^{K-1} \frac{b(h)}{c(h)} \right)^{-1} \prod_{j=1}^{J} \left(\prod_{k=1}^{n_j} \eta(j; x_{jk}) \right) \left(\frac{\prod_{h=1}^{n_j-1} q_j(h)}{\prod_{h=1}^{n_j} p_j(h)} \right)
$$

$$
\cdot \ G(K, J+1)^{-1},
$$

$$
(n_0, (x_{11}, \ldots, x_{Jn_J})) \in \tilde{S}(K, J+1).
$$

Letting $K \longrightarrow \infty$, we have to prove convergence to a finite limit

$$
\lim_{K \to \infty} \left(\prod_{h=0}^{K-1} \frac{b(h)}{c(h)} \right)^{-1} G(K, J+1)^{-1} < \infty.
$$

But this can be proved to be equivalent to (4.1).

Existence of a finite limit implies that for $K \longrightarrow \infty$ the total population size in the transmission line $1, \ldots, J$ remains finite with probability 1. So node 0 is the bottleneck of the system which approaches a state dependent Bernoulli source. $\odot$

Theorem 4.1 opens the way to optimize transmission lines as described in section 4.1 using the principle of adjusted transfer rates and the simple but versatile state dependent arrival process with customers of different types. Allowing furthermore $b(n) \in [0, 1)$ for some $n \in \mathbb{N}$ then theorem 4.1 includes the case of loss systems as well, where at a server with one transmission channel and $L - 1 \geq 0$ buffer places arriving customers are lost when on their arrival the system is full. With suitably restricted state space and adapted norming constant formula (4.2) applies again.(See [31], section 3.)

Example 4.2 (Control of Bernoulli arrival streams) *Consider a state in-dependent Bernoulli arrival stream with intensity $B \in (0,1]$ of customers of different types (an arrival is of type $m \in M$ with probability $a(m)$), which feeds an open tandem of Bernoulli servers as described above. Introduce a Bernoulli switch at the entrance point of the network (before node 1): If the total popula-tion size of the network is n then an arriving customer is allowed to enter with probability $\beta(n) \in (0,1]$ and is rejected and lost with probability $1 - \beta(n)$. As-suming the usual independence properties of a Bernoulli switch this system fits into the class of models where theorem 4.1 is applicable with $b(n) = B \cdot \beta(n)$. If the arrival process is sufficently thin $(1 - \beta(n)$ sufficient high) we have ergodicity. Therefore such a switch can be incorporated to stabilize the system, and we can construct a scheduling regime for optimizing the network.*

Example 4.3 (Open loss systems) *The assumption $\beta(n) \in (0,1]$ on the ad-mission control function $\beta(\cdot)$ can be weakend such that $\beta(n) = 0$ holds if $n \geq L$ for some control limit $L \in \mathbb{N}$. This yields the usual loss control for networks if we have $\beta(n) = 1, 0 \leq n < L$.*

Our aim is to investigate an individual customer's delay behaviour. We the-refore need an arrival theorem similar to proposition 3.5. The proof is similar, but more involved because of the state dependent arrivals. We omit it therefore.

Theorem 4.4 (Arrival Theorem) *Consider the state process X of the open tandem in equilibrium and consider the event*
$A(1,m) = \{$ *at time 0 a customer of type m arrives at node 1*$\}$*. Then*

$$\pi_{1,m}^{J}(x_1,\ldots,x_J) = P(X(0) = ((x_1,m),x_2,\ldots,x_J|A(1,m)) \tag{4.3}$$

$$= P(X(0) = ((x_{11},\ldots,x_{1n_1}),m);(x_{21},\ldots,x_{2n_2});\ldots;(x_{J1},\ldots,x_{Jn_J})|A(1,m))$$

$$= \left(\frac{\prod_{h=0}^{n_1+\cdots+n_J} b(h)}{\prod_{h=0}^{n_1+\cdots+n_J+1} c(h)} \right) \prod_{j=1}^{J} \left(\prod_{k=1}^{n_j} \eta(j;x_{jk}) \right)$$

$$\left(\prod_{h=1}^{n_1} \frac{q_1(h)}{p_1(h)} \right) \prod_{j=2}^{J} \left(\frac{\prod_{h=1}^{n_j-1} q_j(h)}{\prod_{h=1}^{n_j} p_j(h)} \right) \cdot H_1(J)^{-1},$$

$$(x_1,\ldots,x_J) \in \tilde{S}(J),$$

$H_1(J)$ is the norming constant, which does not depend on the type of the arriving customer.
For $i \neq 1$ similar formulas apply.

The usual interpretation of $\pi_{i,m}^{J}$ is that it describes the distribution of the other customers' disposition in an arrival instant at node i under equilibrium con-ditions, the jumping customer is not counted. Remarkable is that this arrival distribution has not the form of the equilibrium of the system. In continuous time such a statement is true. For a state independent Poisson arrival stream at node 1 this is the celebrated PASTA property (Poisson Arrivals See Time Averages) . Another strange observation is that the arrival distributions in discrete time depend on the node, where the arrival appears. This is again contrary to the continuous time case where the arrival disposition has the same distribution for all nodes of the open tandem, [123], [94].

Note: If we rescale time, approaching in the limit a continuous time scale, our systems transform to open exponential queueing networks: In the limit the asymmetry of the arrival distributions disappears and all of them are – at least for state independent arrivals – equal to the steady state distribution of the network. (Similar remarks apply in the closed cycle case of Proposition 3.5.) As we have shown in theorem 3.9 these remarks apply even for closed cycles with state independent service rates and equally it does in case of state independent arrival and service probabilities for open tandems.

Some typical results for providing explicit performance values follow.

Corollary 4.5 (Individual loss probabilities) *(a) Control of Bernoulli arrival streams (Example 4.2): Assume that a Bernoulli arrival process with arrival probability $B \in (0,1]$ is controlled by a Bernoulli switch with admission probabilities $\beta(n), n \in I\!N$. Then the loss probability for an arriving customer of type m due to rejection is*

$$p_{l,m}(J) = 1 - \frac{1}{B \cdot H(J)} \sum_{K=0}^{\infty} \prod_{h=0}^{K} \frac{b(h)}{c(h)} G(K,J),$$

where $G(K,J)$ is the norming constant for the arrival distribution at node 1 for a customer in a closed cycle of J nodes (see section 3.3) with K indistinguishable customers cycling. (See [31]; Theorem 1.)

(b) Open loss system (Example 4.3): Assume additionally that the control of the Bernoulli-(B) process is of the form

$$\beta(n) = 1, \quad n < L, \qquad \beta(n) = 0, \quad n \geq L.$$

Then the loss probability for an arriving customer of type m due to overflow is

$$p_{L,m}(J) = \frac{G(L,J)}{H(J)}.$$

Proof : Similar to the proof of Theorem 4.4. We sketch only part (a).
Let $A(1,m)$ denote the event that at time 0 a customer of type m arrives at node 1, and $A^0(1,m)$ denote the event that at time 0 a customer of type m is

rejected at node 1. We have to compute $P(A^0(1,m)|A(1,m))$. This is

$$\sum_{x \in \tilde{S}(J)} P(\{X-1) = x\} \cap A^0(1,m) \cap A(1,m)) P(A(1,m))^{-1}$$

$$= \sum_{x \in \tilde{S}(J)} \pi^J(x) \cdot (Ba(m))(1 - \beta(n_1 + \cdots + n_J))(Ba(m))^{-1}$$

$$= 1 - H(J)^{-1} \sum_{x \in \tilde{S}(J)} \left(\frac{\prod_{h=0}^{n_1 + \cdots + n_J - 1} b(h)}{\prod_{h=0}^{n_1 + \cdots + n_J} c(h)} \right)$$

$$\cdot \prod_{j=1}^{J} \left(\prod_{k=1}^{n_j} \eta(j; x_{jk}) \left(\frac{\prod_{h=1}^{n_j - 1} q_j(h)}{\prod_{h=1}^{n_j} p_j(h)} \right) \right) \cdot \beta(n_1 + \cdots + n_J)$$

$$\stackrel{(1)}{=} 1 - H(J)^{-1} \sum_{K=0}^{\infty} \frac{1}{B} \left(\prod_{h=0}^{K} \frac{b(h)}{c(h)} \right) \sum_{\substack{(n_1, \ldots, n_J) \in \mathbb{N}^J \\ n_1 + \cdots + n_J = K}} \prod_{j=1}^{J} \left(\frac{\prod_{h=1}^{n_j - 1} q_j(h)}{\prod_{h=1}^{n_j} p_j(h)} \right)$$

$$= 1 - \frac{1}{B \cdot H(J)} \sum_{K=0}^{\infty} \left(\prod_{h=0}^{K} \frac{b(h)}{c(h)} \right) G(K, J).$$

In (1) we have used that $b(n) = B \cdot \beta(n)$ holds. $\odot$

To compute loss probabilities at a single station with finite buffer, single deterministic-(1)-server, and a Markovian arrival stream is transformed in [68] to compute steady state probabilities in an infinite buffer queue. Similar principles can be seen behind the above results on loss probabilities.

Theorem 4.6 (Throughput of the tandem) *In equilibrium the throughput of the tandem is*

$$Th(J) = H_J(J) \cdot H(J)^{-1}.$$

The throughput of type m customers is $a(m) \cdot Th(J)$.

Proof: The throughput of the line is

$$\sum_{(x_1, \ldots, x_J) \in \tilde{S}(J)} \pi^J(x_1, \ldots, x_J) \cdot p_J(n_J).$$

Direct summation yields the result. $\odot$

This result is somewhat curious: Inspection of the definition of $H_j(J)$ in theorem 4.4 leads to the conjecture that the value of the throughput depends on the node where it is evaluated, because the $H_J(J)$ appears. But it can be shown, that $H_j(J)$ does not depend on j. (See lemma 7.3 in section 7.2 of the appendix.)

4.3 Delay Time Distribution for Customers in an Open Tandem of State Independent Bernoulli Servers

To compute the passage time (end–to–end–delay) distribution for the general tandem system of theorem 4.1 with state dependent service times seems to be an unsolved problem up to now. In the following we shall present some partial results for the case of state independent service rates, which is for applications the possibly most important instance.

We consider a test customer of type m arriving at time 0 at node 1 who finds the other customers distributed according to $\pi_{1,m}^J$. Let $(S_1, \ldots, S_J)$ denote the vector of this test customer's successive sojourn times (=waiting time + service time) at the nodes during his passage starting at time 0.

We denote by $P_{\pi_{1,m}^J}$ a probability law which governs the evolution of the system under this initial condition, and by $E_{\pi_{1,m}^J}[\cdot]$ expectations under $P_{\pi_{1,m}^J}$. Working under this initial conditions we have the following theorem [31]:

Theorem 4.7 (Joint sojourn time distribution in a tandem)
Let $(S_1, S_2, \ldots, S_J)$ denote the vector of the test customer's successive sojourn times at the nodes during his passage through the tandem. The joint distribution of $(S_1, S_2, \ldots, S_J)$ is given by the generating function

$$
E_{\pi_{1,m}^J}\left[\prod_{j=1}^J \theta_j^{S_j}\right] = \sum_{(n_1,\ldots,n_J) \in \mathbb{N}^J} H_1(J)^{-1} \tag{4.4}
$$
$$
\cdot \left(\frac{\prod_{h=0}^{n_1+\cdots+n_J} b(h)}{\prod_{h=0}^{n_1+\cdots+n_J+1} c(h)}\right) \left(\frac{q_1}{p_1}\right)^{n_1} \prod_{j=2}^J \left(\frac{1}{q_j}\right)^{\eta(0,n_j)} \left(\frac{q_j}{p_j}\right)^{n_j}
$$
$$
\cdot \left(\frac{p_1\theta_1}{1-q_1\theta_1}\right)^{n_1+1} \prod_{j=2}^J \left\{\left(\frac{p_j\theta_j}{1-q_j\theta_j}\right)^{n_j+1} \left(\frac{1}{\theta_j}\right)^{\eta(0,n_j)}\right\},
$$
$$
|\theta_j| \le 1, j = 1, 2, \ldots, J.
$$

Proof : The arrival probability for the case of state independent service probabilities is

$$
\pi_{1,m}^J(x_1, \ldots, x_J) \tag{4.5}
$$
$$
= P(X(0) = ((x_1, m), x_2, \ldots, x_J | A(1, m))
$$
$$
= \left(\frac{\prod_{h=0}^{n_1+\cdots+n_J} b(h)}{\prod_{h=0}^{n_1+\cdots+n_J+1} c(h)}\right) \prod_{j=1}^J \left(\prod_{k=1}^{n_j} \eta(j; x_{jk})\right)
$$
$$
\cdot \left(\frac{q_1}{p_1}\right)^{n_1} \prod_{j=2}^J \left(\frac{1}{q_j}\right)^{\eta(0,n_j)} \left(\frac{q_j}{p_j}\right)^{n_j} H_1(J)^{-1},
$$
$$
(x_1, \ldots, x_J) \in \tilde{S}(J).
$$

Conditioning on the arrival disposition of the other customers seen by the arriving customer we obtain

$$
E_{\pi_{1,m}^J}\left[\prod_{j=1}^{J}\theta_j^{S_j}\right] \tag{4.6}
$$

$$
= \sum_{(x_1,\ldots,x_J)\in\tilde{S}(J)} \pi_{1,m}^J(x_1,x_2,\ldots,x_J)
$$

$$
\cdot \qquad E_{\pi_{1,m}^J}\left[\prod_{j=1}^{J}\theta_j^{S_j}|X(0)=((x_1,m),x_2,\ldots,x_J)\right]
$$

$$
= \sum_{K=0}^{\infty} H_1(J)^{-1}\left(\frac{\prod_{h=0}^{K}b(h)}{\prod_{h=0}^{K+1}c(h)}\right)G_{1,m}(K+1,J)
$$

$$
\cdot \left\{\sum_{\substack{(x_1,\ldots,x_J)\in\tilde{S}(J)\\ n_1+\cdots+n_J=K}} E_{\pi_{1,m}^J}\left[\prod_{j=1}^{J}\theta_j^{S_j}|X(0)=((x_1,m),x_2,\ldots,x_J)\right]\right.
$$

$$
\left. \cdot \prod_{j=1}^{J}\left(\prod_{k=1}^{n_j}\eta(j;x_{jk})\right)\left(\frac{q_1}{p_1}\right)^{n_1}\prod_{j=2}^{J}\left(\frac{1}{q_j}\right)^{\eta(0,n_j)}\left(\frac{q_j}{p_j}\right)^{n_j}G_{1,m}(K+1,J)^{-1}\right\}.
$$

The sum in the waved brackets can be given a specific interpretation:

Consider a closed cycle as described in section 3.1 with $K+1$ customers cycling, and with a customer of type m, henceforth called C_m, on arrival at node 1, seeing the other customers distributed according to the arrival distribution of proposition 3.5. In case of state independent service probabilities this distribution is

$$
\pi_{1,m}^{K+1,J}(x_1,\ldots,x_J) \tag{4.7}
$$

$$
= \prod_{j=1}^{J}\left(\prod_{k=1}^{n_j}\eta(j;x_{jk})\right)\left(\frac{q_1}{p_1}\right)^{n_1}\prod_{j=2}^{J}\left(\frac{1}{q_j}\right)^{\eta(0,n_j)}\left(\frac{q_j}{p_j}\right)^{n_j}G_{1,m}(K+1,J)^{-1},
$$

$$
(x_1,\ldots,x_J)\in\tilde{S}(K,J).
$$

Now we consider in this virtual closed cyclic system the joint conditional distribution of C_m's successive sojourn times during his next cycle, given the arrival disposition $(x_1,\ldots,x_J)$ for the other customers and conclude that it is identical to the joint conditional distribution of a customer on his passage through the open tandem, given that customer observes an arrival disposition $(x_1,\ldots,x_J)$. This is due to the strong Markov property of the state processes for both systems, the identical structure of the nodes in the systems, and the *overtake–free property* of the open tandem as well as of the closed cyclic network with respect to *one* cycle of C_m.

We therefore can reduce the computation of (4.6) to first computing the joint sojourn time distributions of C_m in a sequence of closed networks with increasing population size $K + 1 = 1, 2, \ldots$. Using the obvious notation similar to (4.3) and (4.6) we have to compute

$$E_{\pi_{1,m}^{K+1,J}}\left[\prod_{j=1}^{J}\theta_j^{S_j}\right]$$

$$= \left\{\sum_{\substack{(x_1,\ldots,x_J)\in\tilde{S}(J)\\n_1+\cdots+n_J=K}} E_{\pi_{1,m}^{K+1,J}}\left[\prod_{j=1}^{J}\theta_j^{S_j}|X(0) = ((x_1,m), x_2,\ldots,x_J)\right]\right. \tag{4.8}$$

$$\cdot \left. \prod_{j=1}^{J}\left(\prod_{k=1}^{n_j}\eta(j; x_{jk})\right)\left(\frac{q_1}{p_1}\right)^{n_1}\prod_{j=2}^{J}\left(\frac{1}{q_j}\right)^{\eta(0,n_j)}\left(\frac{q_j}{p_j}\right)^{n_j}G_{i,m}(K+1,J)^{-1}\right\}.$$

Collecting states in (4.8) according to common joint queue length vectors and arguing that the conditional sojourn time distribution does not depend on the types of the customers we can apply the result of theorem 3.11 and obtain

$$E_{\pi_{1,m}^{K+1J}}\left[\prod_{j=1}^{J}\theta_j^{S_j}\right] \tag{4.9}$$

$$= \sum_{\substack{(n_1,\ldots,n_J)\in\mathbb{N}^J\\n_1+\cdots+n_J=K}} G_1(K+1,J)^{-1}\left(\frac{q_1}{p_1}\right)^{n_1}\prod_{j=2}^{J}\left(\frac{1}{q_j}\right)^{\eta(0,n_j)}\left(\frac{q_j}{p_j}\right)^{n_j}$$

$$\cdot \left(\frac{p_1\theta_1}{1-q_1\theta_1}\right)^{n_1+1}\prod_{j=2}^{J}\left\{\left(\frac{p_j\theta_j}{1-q_j\theta_j}\right)^{n_j+1}\left(\frac{1}{\theta_j}\right)^{\eta(0,n_j)}\right\},$$

$$|\theta_j| \le 1, j = 1, 2, \ldots, J.$$

Inserting this into (4.6) yields (4.4). $\odot$

The reader should be cautioned: The tempting conjecture

$$E_{\pi_{1,m}^J}\left[\prod_{j=1}^{J}\theta_j^{S_j}|X(0) = ((x_1,m), x_2,\ldots,x_J)\right]$$

$$= \left(\frac{p_1\theta_1}{1-q_1\theta_1}\right)^{n_1+1}\prod_{j=2}^{J}\left\{\left(\frac{p_j\theta_j}{1-q_j\theta_j}\right)^{n_j+1}\left(\frac{1}{\theta_j}\right)^{\eta(0,n_j)}\right\}$$

$$= E_{\pi_{1,m}^{K+1,J}}\left[\prod_{j=1}^{J}\theta_j^{S_j}|X(0) = ((x_1,m), x_2,\ldots,x_J)\right]$$

is false. This is the reason why the proofs of (4.9) and (4.4) are difficult. Compare the discussion following theorem 3.11.

The result of theorem 4.9 states that the joint distribution of the successive sojourn times of a customer in equilibrium is distributed like a mixture of multivariate distributions with independent negative binomial marginals. The mixture distribution is explicitely given. So the inversion of the transform result (4.6) can be performed easily by hand. These remarks apply to the following results as well. From (4.4) we obtain the distribution of a customer's end–to–end–delay $S_1 + \cdots + S_J$.

Corollary 4.8 (End–to–end–delay distribution) *For the test customer of type m arriving at time 0 at node 1, finding the other customers distributed according to $\pi_{1,m}^J$, the distribution of the end–to–end–delay is given by the generating function*

$$E_{\pi_{1,m}^J}\left[\theta^{S_1+\cdots+S_J}\right]$$

$$= \sum_{(n_1,\ldots,n_J)\in I\!N^J} H_1(J)^{-1}$$

$$\left(\frac{\prod_{h=0}^{n_1+\cdots+n_J} b(h)}{\prod_{h=0}^{n_1+\cdots+n_J+1} c(h)}\right)\left(\frac{q_1}{p_1}\right)^{n_1} \prod_{j=2}^{J}\left(\frac{1}{q_j}\right)^{\eta(0,n_j)}\left(\frac{q_j}{p_j}\right)^{n_j}.$$

$$\left(\frac{p_1\theta}{1-q_1\theta}\right)^{n_1+1} \prod_{j=2}^{J}\left\{\left(\frac{p_j\theta}{1-q_j\theta}\right)^{n_j+1}\left(\frac{1}{\theta}\right)^{\eta(0,n_j)}\right\}, \quad |\theta|\leq 1.$$

For the case of a state independent Bernoulli arrival process these results boil down to simply to evaluate distributions. The most remarkable result in this case is the independence of the successive sojourn times of a customer, when traversing the tandem [29].

Corollary 4.9 (Independent sojourn times) *For state independent Bernoulli arrivals the joint distribution of $(S_1, S_2, \ldots, S_J)$ is given by the generating function*

$$E_{\pi_{1,m}^J}\left[\prod_{j=1}^{J}\theta_j^{S_j}\right] = \prod_{j=1}^{J}\frac{\left(\frac{p_j-b}{c}\right)\theta_j}{1-\left(1-\frac{p_j-b}{c}\right)\theta_j}, \quad |\theta_j|\leq 1, j=1,2,\ldots,J.$$

The individual sojourn times are geometrically distributed with parameter $\frac{p_j-b}{c}$, $j=1,\ldots,J$, and independent.

This result is the analogue of Burke's and Reich's results on the independence of a customer's successive sojourn times in an open exponential tandem in continuous time (for a review see [9]). An immediate consequence is that the distribution of a customer's end-to-end-delay in a network in equilibrium can be computed directly as a convolution of the single-node-delay distributions, for a direct proof see [26]:

Corollary 4.10 (End–to–end–delay) *For the network in equilibrium the transition time (end–to–end–delay) of a customer is given by*

$$E\left[\theta^{S_1+\cdots+S_J}\right] = \prod_{j=1}^{J} \frac{(\frac{p_j-b}{c})\theta}{1-(1-\frac{p_j-b}{c})\theta} \quad , \quad |\theta| \leq 1. \tag{4.10}$$

A typical example of an adhoc approximation procedure to solve complex discrete time network problems is described by Bruneel and Kim [14], Chapter 4.1.6: The end–to–end delay of a cell on its transmission through an ATM network is computed by *assuming that the successive single node delays behave statistically independent.* This results in convolution formulas for the end–to–end delay distribution. The result on the end–to–end delay in corollary 4.10 is therefore: In case of linear series of state independent Bernoulli servers *the assumption holds* and the convolution formula is *exact.* Clearly, compared with the decomposition approximation of Bruneel and Kim, formula (4.10) suffers from the fact that the class of networks dealt with is much more narrow.

The approach of Bruneel and Kim can be viewed as typical for dealing with more general cases. Our corollaries 4.9, 4.10 contribute to the discussion in that we have identified some fundamental networks where their approximation is exact.

Similar procedures and results are well known in continuous time networks. The decompsosition of the end–to–end–delay into independent sojourn times as an approximation procedure is (e.g.) investigated in [64] and termed IFTA there *(independent flow time analysis).* For overtake free paths it is known that the IFTA procedures yields exact distributional results. For a discussion see [9], sections II.7 and III.1.

4.4 Re–entrant Lines

Re–entrant lines are models for manufacturing systems of high complexity with items (parts) flowing through the line which possibly request for different kind of service and different amount of work. At control stages in the line after a production step the items are controlled and according to the observed quality of production either released to the next step in the process or send back to the previous process node for rework. Examples for such systems are semiconductor manufacturing plants. Typically these production networks show a linear structure, which is superimposed by feedback cycles with reentering machines more than once. For a short introduction into the modeling of re–entrant lines via queueing networks see [90], and the references there. Scheduling of such systems is described e.g. in [91].

If the actions on all stages are synchronized, then the production line is suitably modeled by a discrete time network of queues. A more detailed description may be found in examples 12.1 and 12.37 of [18]. The models considered in this section revisit these examples. It should be remarked that the re–entrant systems considered here do not meet the full generality which is inherent in the formulation of re–entrant lines. We only allow immediate return to the same machine, which is necessary for the manufacturing systems with production control, but it is not sufficient to model parts visiting machines more than once in course of the planned production process. This problem is almost unsolved in the framework considered here and is part of ongoing research in discrete time stochastic networks.

For the more versatile framework in continuous time which includes such behaviour see the tutorial [90]. For the discrete time networks a first attempt is described in section 5.3, and [33], respectively. More recent developments are described in section 6.1. Note however that the models in section 6.1 do not explicitly model part specific sequencing in the nodes. This can be easily incorporated into the framework presented here by introducing different customer types.

Rework of an item is modeled by immediate feedback of the item into the same node. So the feedback nodes of section 2.3 will be suitable models for the stages in the production process. For ease of presentation we shall restrict the description to the case of indistinguishable customers, i.e. the products which flow through the line are homogeneous.

Our first process model is a system which is assumed to be able to handle only a limited number of at K items in parallel. We assume that there is heavy load for the system. This means that at every time epoch when an item eventually leaves the line, departing successfully without rework from its final stage, a new item is fed into the entrance node of the line. Consequently the production line is considered as a closed line, i.e. a cycle of nodes with immediate feedback. We combine the cyclic queue of section 3.1 with the feedback node structure from section 2.3.

We allow the production capacity of a nodes to be adjusted to the load of that node: the service capacity is state dependent. The reliability of the production depends on the load of the node, the feedback probability is therefore state dependent as well. These modeling demands are met by the Bernoulli server with immediate feedback of section 2.3.

The details: We consider a cycle of J state dependent Bernoulli servers with immediate feedback under FCFS as described in section 2.3 with K indistinguishable customers cycling. A customer in service at node j (where the population size is n_j) departes from that node with probability $p_j(n_j)$, leaving behind $n_j - 1$ customers. He is fed back into the waiting room (to the tail of queue j) with probability $r_j(n_j)$. If he was the only customer present he will obtain immediately a further service, otherwise he will join the tail of the queue. With probability $1 - r(n_j)$ he will leave the node and proceed to the next node $j+1$ of the system ($J + 1 := 1$). The decision whether to complete the actual service at node j and whether to leave or to reenter the node is made independently of anything else in the history of the system given the actual queue length of that node. The regulation of customer movements in case of multiple events is: Departure before arrival for a joint arrival and departure (D/A) and feedback before arrival for a joint feedback and arrival event (F/A). This regulation scheme synchronizes the movements of customers all over the system.

The evolution of the system will be described by a multivariate discrete time stochastic process $X := (X(t) : t \in I\!N)$ as follows:

Let $X_j(t)$ denote the queue length (customers in service + customers waiting) at node j at time t. Then $X(t) = (X_1(t), \ldots, X_J(t))$ is the joint queue length process on state space

$$\tilde{S}(K, J) := \{(n_1, \ldots, n_J) \in N^J : n_1 + \cdots + n_J = K\},$$

which is sufficient for X being Markov, irreducible on $\tilde{S}(K, J)$, aperiodic, and positive recurrent with unique limiting and stationary distribution.

Theorem 4.11 (Steady state of the cycle of feedback queues)
Let $X = (X_t : t \in I\!N)$ denote the joint queue length process of the cycle of state dependent Bernoulli servers with immediate feedback. The unique stationary and limiting distribution for X is $\pi^{K,J} = (\pi^{K,J}(n_1, \ldots, n_J) : (n_1, \ldots, n_J) \in \tilde{S}(K, J))$ with

$$\pi^{K,J}(n_1, \ldots, n_J) \;=\; \prod_{j=1}^{J} \frac{\prod_{h=1}^{n_j-1}(q_j(h) + p_j(h)r_j(h))}{\prod_{h=1}^{n_j} p_j(h)(1 - r_j(h))} \cdot G(K, J)^{-1},$$

$$(n_1, \ldots, n_J) \in \tilde{S}(K, J), \tag{4.11}$$

where $G(K, J)$ is the norming constant.

Similar to proposition 3.3 we can identify easily a system which is described by the time reversal of X.

Proposition 4.12 (Time Reversed Chain) *Denote by $X = (X(t) : t \in Z\!\!\!Z)$ the stationary continuation of X under $\pi^{K,J}$ on time axis $Z\!\!\!Z$ as well. Then the time reversal of X*

$$\bar{X} = (\bar{X}(t) := X(-t) : t \in Z\!\!\!Z),$$

describes a closed cycle of state–dependent single server nodes with immediate feedback. The node characteristics are the same as those of the original system but with K customers cycling in the reversed direction. For state $n = (n_1, \ldots, n_J) \in \tilde{S}(K, J)$ of $\bar{X}$ the local states n_j denote local queue lengths of the nodes.

Proof of theorem 4.11 and proposition 4.12: Let $p = (p(m, n) : m, n \in \tilde{S}(K, J))$ denote the transition matrix of X and $\bar{p} = (\bar{p}(m, n) : m, n \in \tilde{S}(K, J))$ the transition matrix of $\bar{X}$. For any $m, n \in \tilde{S}(K, J)$ we have to show (see proposition 7.2 in the appendix.)

$$\pi^{K,J}(m)p(m, n) = \pi^{K,J}(n)\bar{p}(n, m). \tag{4.12}$$

Let for $m = (m_1, \ldots, m_J) \in \tilde{S}(K; J)$ denote $A(m) := \{j \in \{1, \ldots, J\} : m_j > 0\}$ the set of busy nodes when the network is in state m. One–step successor states $n = (n_1, \ldots, n_J) \in \tilde{S}(K, J)$ of m, $n \neq m$, can be determined as follows and are selected according to the following rules and probabilities:

Chose some $B \subseteq A(x), B \neq \phi, B \neq \{1, \ldots, J\}$, with probability

$$\prod_{j \in B} p_j(m_j)(1 - r_j(m_j)) \cdot \prod_{j \in A(x) - B} (q_j(m_j) + p_j(m_j)r_j(m_j)).$$

At nodes in B a service will expire at the end of the present time slot and the customer will not return to node j but immediately proceed to node $j+1$. Denote the successor of m obtained in this way by n. It follows

$$n_j = \begin{cases} m_j + 1, & \text{if} \quad j \notin B, \quad j - 1 \in B \\ m_j - 1, & \text{if} \quad j \in B, \quad j - 1 \notin B \\ m_j, & \text{if} \quad (j \in B, \quad j - 1 \in B) \quad \text{or} \quad (j \notin B, \quad j - 1 \notin B), \end{cases}$$

for $j = 1, \ldots, J$.

For $B \subseteq A(x), B \neq \phi, B \neq \{1, \ldots, J\}$, define
$E(B) = \{j \in \{1, \ldots, J\} : j \in B, j - 1 \notin B\},$ (the END of B)
and
$F(B) = \{j \in \{1, \ldots, J\} : j - 1 \in B, j \notin B\},$ (in FRONT of B).

Note that for each pair $M \neq n$ of successor states there exists exactly one nonempty set $B \subseteq A(m)$ such that by departing of customers from exactly the nodes in B results in a transition from m to n, and vice versa. Then (4.12) reads

$$\prod_{j=1}^{J} \frac{\prod_{h=1}^{m_j-1}(q_j(h) + p_j(h)r_j(h))}{\prod_{h=1}^{m_j} p_j(h)(1 - r_j(h))}$$

$$\cdot \prod_{j \in B} p_j(m_j)(1 - r_j(m_j)) \cdot \prod_{j \in A(x) - B} (q_j(m_j) + p_j(m_j)r_j(m_j))$$

$$= \prod_{j \in F(B)} \frac{\prod_{h=1}^{m_j}(q_j(h) + p_j(h)r_j(h))}{\prod_{h=1}^{m_j+1} p_j(h)(1 - r_j(h))} \cdot \prod_{j \in E(B)} \frac{\prod_{h=1}^{m_j-2}(q_j(h) + p_j(h)r_j(h))}{\prod_{h=1}^{m_j-1} p_j(h)(1 - r_j(h))}$$

$$\cdot \prod_{j \in (\{1,\ldots,J\} - E(B) - F(B))} \frac{\prod_{h=1}^{m_j-1}(q_j(h) + p_j(h)r_j(h))}{\prod_{h=1}^{m_j} p_j(h)(1 - r_j(h))}$$

$$\cdot \prod_{j \in F(B)} p_j(m_j + 1)(1 - r_j(m_j + 1)) \cdot \prod_{j \in B - E(B)} p_j(m_j)(1 - r_j(m_j))$$

$$\cdot \prod_{j \in E(B)} (q_j(m_j - 1) + p_j(m_j - 1)r_j(m_j - 1)))^{\eta(1,m_j)}$$

$$\cdot \prod_{j \in A(x) - B - F(B)} (q_j(m_j) + p_j(m_j)r_j(m_j))$$

This can be checked directly. $\odot$

Next we consider an open re–entrant line, i.e., an open tandem similar to the models in section 4.2 with nodes of Bernoulli servers with immediate feedback. The details of the system are:

The node structure and the feedback mechanism is the same as before in the closed cycle: state dependent service probabilities and feedback probabilities. The arrival probabilities for the (undistinguishable) customers are $b(n) \in (0,1)$ if the total population size of the system is n. Similar to the reasoning before theorem 4.11 we conclude: The evolution of the system can be described by a multivariate discrete time Markov chain $X := (X(t) : t \in I\!N)$ as follows:

$X_j(t)$ is the queue length (customers in service + customers waiting) at node j at time t, $X(t) = (X_1(t), \ldots, X_J(t))$ is the joint queue length process, irreducuble on state space $I\!N^J$, and aperiodic. If X is positive recurrent we obtain a unique limiting and stationary distribution, which provides us with conditions for stabilizing the re–entrant line.

Theorem 4.13 (Steady state of the tandem of feedback queues)
$X = (X_t : t \in I\!N)$, *the joint queue length process of the open tandem of state dependent Bernoulli servers with immediate feedback is positive recurrent, hence ergodic, if and only if*

$$H(J) := \sum_{(n_1,\ldots,n_J)\in I\!N^J} \frac{\prod_{h=0}^{n_1+\cdots+n_J-1} b(h)}{\prod_{h=0}^{n_1+\cdots+n_J} c(h)} \prod_{j=1}^{J)} \frac{\prod_{h=1}^{n_j-1}(q_j(h) + p_j(h)r_j(h))}{\prod_{h=1}^{n_j} p_j(h)(1 - r_j(h))} < \infty.$$

(4.13)

If X is ergodic, its unique stationary and limiting distribution is
$\pi^J = (\pi^J(n_1,\ldots,n_J) : (n_1,\ldots,n_J) \in I\!N^J)$ *with*

$$\pi^J(n_1,\ldots,n_J)$$

(4.14)

$$= \frac{\prod_{h=0}^{n_1+\cdots+n_J-1} b(h)}{\prod_{h=0}^{n_1+\cdots+n_J} c(h)} \prod_{j=1}^{J)} \frac{\prod_{h=1}^{n_j-1}(q_j(h) + p_j(h)r_j(h))}{\prod_{h=1}^{n_j} p_j(h)(1 - r_j(h))} \cdot H(J)^{-1},$$

$$(n_1,\ldots,n_J) \in I\!N^J,$$

(4.15)

where $H(J)$ is the norming constant.

The proof can be performed simililarly to that of theorem 4.1. An extra node without feedback is introduced which simulates the state dependent Bernoulli source and then the population size is increased to infinity.

The proof is easier for the case of state independent arrival probabilities. Then it can be shown that the single feedback node of corollary 2.17 in equilibrium is reversible according to the proof of Hsu and Burke [65] and quasi reversible according to definition 12.6 of Chao, Miyazawa, and Pinedo [18]. Note however, that the departure stream leaving the server and including the customers to be fed back is not a Bernoulli process. Only the departure process of customers leaving the node is Bernoulli.

It follows that the nodes can be coupled directly to a tandem with product form
steady state.

Theorem 4.14 (Tandem of feedback queues with Bernoulli arrivals)
*Let $X = (X_t : t \in \mathbb{N})$ be the joint queue length process of the open tandem of
state dependent Bernoulli servers with immediate feedback, fed by a state independent Bernoulli(b) process. X is positive recurrent, hence ergodic if and only
if*

$$H(J) := \sum_{(n_1 \ldots, n_J) \in \mathbb{N}^J} \frac{b^{\max(n_1 + \cdots + n_J - 1, 0)}}{c^{n_1 + \cdots + n_J}} \cdot \prod_{j=1}^{J} \frac{\prod_{h=1}^{n_j - 1}(q_j(h) + p_j(h)r_j(h))}{\prod_{h=1}^{n_j} p_j(h)(1 - r_j(h))} < \infty.$$

$$(4.16)$$

*If X is ergodic, the unique stationary and limiting distribution is
$\pi^J = (\pi^J(n_1, \ldots, n_J) : (n_1, \ldots, n_J) \in \mathbb{N}^J)$ with*

$$\pi^J(n_1, \ldots, n_J) \qquad\qquad\qquad\qquad\qquad\qquad (4.17)$$

$$= \frac{b^{\max(n_1 + \cdots + n_J - 1, 0)}}{c^{n_1 + \cdots + n_J}} \prod_{j=1}^{J} \frac{\prod_{h=1}^{n_j - 1}(q_j(h) + p_j(h)r_j(h))}{\prod_{h=1}^{n_j} p_j(h)(1 - r_j(h))} \cdot H(J)^{-1},$$

$$(n_1, \ldots, n_J) \in \mathbb{N}^J, \qquad\qquad\qquad\qquad\qquad\qquad (4.18)$$

where $H(J)$ is the norming constant.

Remark 4.15 (Entrance control and loss systems) *Similar to example 4.2
the result of theorem 4.13 can be generalized by redefining $b(\cdot)$ to include the
case of an open re–entrant line with entrance control. Especially the analogue of
example 4.3 then yields a re–entrant line which has a maximal total capacity. The
heavy traffic assumption of theorem 4.11 is not necessary for this modification.*

Chapter 5

Networks with Doubly Stochastic and Geometrical Servers

The doubly stochastic server was introduced by Schassberger [117] as a discrete time analogue of Kelly's *symmetric server* [75]. The latter is a generalization and unification of the nodes which are used to build the BCMP networks in continuous time [4]: Nodes with processor sharing (PS) or Last–Come–First–Served–preemptive resume (LCFS–preemptive resume) discipline or infinite servers. Baskett, Chandy, Muntz, and Palacios showed that customers may request at these nodes for type dependent non exponential service times and nevertheless a generic product form steady state is found.

The exponential single servers under FCFS which are further possible building blocks of the BCMP networks are generalized by Kelly as well to include *general exponential nodes*: Non symmetric servers with customer type independent exponential service request. The discrete time analogue of these nodes are the geometrical nodes to be described below.

Kelly's networks are a generalisation of the BCMP networks, because he removed the restriction to only *PS, LCFS–preemptive resume, infinite server, and exponential FCFS* regimes for the nodes by introducing the *general exponential* and *symmetric servers*. For these networks the steady state probabilities can be explicitly given in simple terms as in the BCMP case and the main performance quantities can be computed as well. An appealing property is that the first order mean values of the relevant performance measures are insensitive: They remain invariant under variation of the customers' service time distribution at symmetric servers as long as the mean service time remains invariant.

We shall present in this chapter a class of discrete time networks with general network topology which resemble the BCMP and Kelly networks and sketch some of their properties. The counterpart to the symmetric server with non exponential service time request is doubly stochastic server described in section 5.1,

H. Daduna: Queueing Networks with Discrete Time Scale, LNCS 2046, pp. 73–103, 2001.

the conterpart of the exponential (multiserver) node is the geometrical server in section 5.2. We describe their steady state behaviour, which is of an internal product form for the nodes. Open networks with doubly stochastic and geometrical nodes and deterministic routing of customers are described in section 5.3, the steady state is presented without proof in section 5.4, it is of external product form over the nodes of the network. In section 5.5 we derive explicit closed form expressions for important performance measures of the network. The parallel closed network case is dealt with in section 5.6, including examples for computational algorithms.

In section 5.7 we specialize the network's topology to the case of open tandems. It is shown that the rules for regulation of simultaneous events, imposed on the network in the general case, can be weakened in this case considerably for an important subclass of nodes.

5.1 Description of the Doubly Stochastic Server

We consider in this section a single node which is fed by a state dependent arrival stream of customers of types $m \in M$, M being a countable set of types (not containing natural numbers for making the later state description unambiguous).

The node consists of an unlimited sequence of service and waiting positions 1,2,3, ... which is controlled according to the *shift–protocol*. Whenever there are n customers present, $n \geq 1$, they occupy positions $1, \ldots, n$. If the customer in position $i \in \{1, \ldots, n\}$ departs from the node then the gap is closed by shifting the customers previously on positions $i + 1, \ldots, n$ one step down into positions $i, \ldots, n - 1$, leaving their internal order invariant. Likewise if an additional customer is inserted into position $i \in \{1, \ldots, n\}$, the customers previously on positions $i, \ldots, n$ are shifted one step up into positions $i + 1, \ldots, n + 1$, leaving their internal order invariant. (If $n = 0$, the node is *empty*.)

Having $n > 0$ customers present at the node service is provided to those customers staying on positions $1, \ldots, C(n)$, where $C(n) > 0$ is a node specific service parameter. We then call positions $1, \ldots, C(n)$ *busy*, while positions $C(n) + 1, \ldots, n$ are said to be *idle*. (We set $C(0) := 0$.)

Customers entering the node request for a certain amount of service time which will be provided to them according to the service disciplin (to be described below) of the node.

If at time $t \in I\!N$ there are n customers present then there will be no arrival with probability $c(n) = (1 - b(n))$ – or there is exactly one arrival which is of type m with probability $b(n) \cdot a(m) > 0$, $m \in M$, such that $\sum_{m \in M} a(m) = 1$ holds.

We assume the successive arrival decisions to depend on the history of the system only through the total population size in system at the begin of the present time slot and the type decisions to be independent of the system's history. We assume late arrivals and for multiple events we assume the D/A rule (departure before arrival), for details see the description before theorem 2.3. The state of the system is recorded at times $t \in \mathbb{N}$, just after possible departures and arrivals have happened.

The amount of service a customer requests for depends deterministically on the customer's type. A customer of type $m \in M$ will request for an amount of $K(m)$ time units of service time, $K(m) \in \mathbb{N}_+$. To exclude trivialities we assume that there exists at least one customer type who requests for more than one time unit of service time.

The requirement of having deterministic service times is not a restriction but opens the possibility of versatile modeling probabilistic behaviour of customers. The key is the introduction of different customer types and type dependent behaviour: The random decision for the service requests are done by selecting the customer's type suitably when entering the network.

To be more specific: If we start with a set of (physical) customer types requesting for service according to general type dependent distributions we may discriminate between sampled different requests for a given customer type by introducing ficticious customer types. Each ficticious customer type carries information about the physical type of that customer, and his exact amount of requested service. This concept will allow stochastically dependent service requests at the successive *doubly stochastic* nodes on a customer's itinerary in the networks defined in sectioon 5.3 and 5.6.

Following from this description the construction of a suitable state space for a Markovian description of the node's evolution necessarily must include at least the following state information – from the description of the service disciplines below it will follow that this information does already suffice:
The state space S consists of elements x as follows:

$x := e$ for the empty node, and sequences

$x := [m(n), k(n); \ldots ; m(1), k(1)], \quad n \geq 1$

where n is the number of customers present at the node (*queue length of the node*), $m(i)$ is the type of the customer on position i, and $k(i)$ his residual request for service time (his *residual work*).

Roughly the service disciplines can be now be described by the following four–step procedure, where we assume throughout $x := [m(n), k(n); \ldots ; m(1), k(1)]$ or $x = e$ to be a generic state as described above:

Definition 5.1 (Doubly stochastic discipline. Part I)

(1) A customer present at time t in a busy position $i \in \{1, \ldots, C(n)\}$ obtains exactly one unit of service time until time $t + 1$.
If $k(i) > 1$ then at time $(t+1)-$ this residual workload is diminished to $k(i) - 1$. If $k(i) = 1$ then this customer departs from the node at time $(t+1)-$. (But notice the restriction on the departure rules given below in (4).)

(2) Assume a customer of type m observes just before his entrance (which will happen between $t-$ and t) the node in state x.
If $x = e$, then the state changes to

$$[m, K(m)];$$

else if $x := [m(n), k(n); \ldots; m(1), k(1)]$ is the state of the system if at time $t-$ all residual service times of customers on busy positions are decreased, and if no arrival appeared,
then with probability $1/C(n+1)$ the state changes to

$$x := [m(n), k(n); \ldots; m(i+1), k(i+1); m, K(m); m(i-1), k(i-1); \ldots; m(1), k(1)],$$

for some $i \in \{1, \ldots, C(n+1)\}$. A new arrival is inserted with probability 1 into a busy position and therefore has immediate access *to service capacity. His position is randomly selected among the busy positions.*

(3) Due to the discrete time scale at any time point there may occur multiple events *composed of several arrival and departure events. Our rule to handle these multiple events resembles what is known as* rejection blocking *or* repetitive service *in multiple access transmission systems with limited buffer capacity [105], p. 455. The details:*

If a multiple event occurs where several customers simultaneously wish to enter and/or to leave nodes of the network, then not all of the requested transitions are allowed:

Customers wishing jointly to depart due to their service completions have to stay at their present position for obtaining just another service (retrial of transmission). This request for service time is deterministically selected and identical to the previous service there. The only exception is:

If several services expire jointly and at the same time instant an arrival occurs then from the departure candidates we select randomly one who departs and is substituted at his position by the new arrival.

Therefore just one single external arrival and at most one service completion with a subsequent departure from the node can be observed. Especially no access conflicts with respect to the handling of multiple events can happen.

(4) Assume that according to (1) or (2) exactly one arrival or one service completion or according to (3) a multiple event appeared and is handled. Then all the stay–on customers, i.e., those customers staying before on idle positions and customers on busy positions whose service did not expire, are permuted on their positions according to some probability law. This law may depend on what has happend in (1),(2),(3) and on the state of the node. (For more details see the description in part II, definition 5.2 and part III, definition 5.5.)

The only restriction is: Given an event as described in (1),(2),(3), then the transition matrix which governs the feasible state transtions due to the permutations is in a certain way doubly stochastic. *(See below, Part III of the definition.) If this is the case the node is called* doubly stochastic *or a* doubly stochastic server. *Such nodes are said to work under a* doubly stochastic discipline.

We now give the details of the *doubly stochastic disciplines* according to [117]. We do not use that the transition laws exhibit the property of being doubly stochastic. Therefore we define a more general discipline but shall finally reduce our investigations to the case of doubly stochastic systems to obtain an explicit product form steady state.

In the following for a given state $x \in S$ we shall call customers to be *stay–on customers* if they either occupy an idle position or if they are on a busy position $i \leq C(n)$ showing a residual work of $k(i) > 1$. A customer is a *departure candidate* if he is in a busy position i with residual work $k(i) = 1$. Customers on some position i are called a *fresh customer* if they show their total service request $K(m) = k(i)$ as residual work.

Definition 5.2 (Doubly stochastic discipline. Part II)

(A) If the state of the node at time t is $x := [m(n), k(n); \ldots; m(1), k(1)] \neq e$, and if there is no departure candidate and no arrival occurs at time $t+1-$, then the set $A(x)$ of possible successor states at time $t+1$ of x is obtained as follows:

Decrease the residual work of customers in busy positions by one and than permute the positions of the customers according to any permutation resulting in state $y \in A(x)$. This happens with probability

$$c(n)d(x,y) \geq 0, \quad , \quad \sum_{y \in A(x)} d(x,y) = 1$$

.

(B) If the state of the node at time t is $x := [m(n), k(n); \ldots; m(1), k(1)]$ or $x = e$, and if there is no departure candidate and an arrival of type $m \in M$ occurs at time $t+1-$, then the set $A_{m,i}(x)$ of possible successor states of x with the new arrival inserted in position i, $i \in 1, \ldots, C(n+1)$, at time $t+1$ is obtained as follows:

Decrease the residual work of customers in busy positions by one, insert the new arrival in position i and than permute the stay–on customers on positions $1, \ldots, i-1, i+1, \ldots, n+1$ according to any permutation resulting in state $y \in A_{m,i}(x)$. This happens with probability

$$b(n)a(m)C(n+1)^{-1}d^+(x,y) \geq 0, \quad \sum_{y \in A_{m,i}(x)} d^+(x,y) = 1.$$

(Note that a new arrival is taken into service immediately, his position is chosen at random, and the rearrangenment of the stay–on customers may depend on the previous state of the node and on the type and position of the new arrival.)

(C) If the state of the node at time t is $x := [m(n), k(n); \ldots; m(1), k(1)] \neq e$, and if there is exactly one depature candidate and no arrival occurs at time $t+1-$, then the set $A(x)$ of possible successor states at time $t+1$ of x is obtained as follows:

Decrease the residual work of customers in busy positions by one, delete the departure candidate and than permute on positions $1, \ldots, n-1$ the stay–on customers according to any permutation resulting in state $y \in A(x)$. This happens with probability

$$c(n)d^-(x,y) \geq 0, \quad \sum_{y \in A(x)} d^-(x,y) = 1.$$

(Note that the rearrangenment of the stay–on customers may depend on the pre-vious state of the node and on the type and position of the departing customer.)

(D) *If the state of the node at time t is $x := [m(n), k(n); \ldots; m(1), k(1)] \neq e$, and if there are $k \geq 1$ departure candidates occupying positions $i_1, \ldots, i_k$, and an arrival of type $m \in M$ occurs at time $t + 1-$, then the set $A_{m,i}(x)$ of possible successor states of x with the new arrival staying in i at time $t + 1$, $i \in 1, \ldots, C(n) + 1$, is obtained as follows:*

Decrease the residual work of customers in busy positions by one, select at random one of the departure candidates who is allowed to depart and the new arrival is inserted in his previous position. All other departure candidates stay on their possitions and become fresh jobs again requesting for a further service of $K(m(i_l))$ time units, $l = 1, \ldots, k$. Then permute the stay–on customers on positions $\{1, \ldots, n\} - \{i_1, \ldots, i_k\}$ according to any permutation resulting in state $y \in A_{m,i}(x)$. This happens with probability

$$b(n)a(m)k^{-1}d^+(x, y) \geq 0, \quad , \quad \sum_{y \in A_{m,i}(x)} d^+(x, y) = 1.$$

(Note that the rearrangement of the stay–on customers may depend on the previous state of the node and on the type and position of the new arrival.)

(E) *If the state of the node at time t is $x := [m(n), k(n); \ldots; m(1), k(1)] \neq e$, and if there are $k \geq 2$ departure candidates occupying positions $i_1, \ldots, i_k$, and no arrival occurs at time $t + 1-$, then the set $A(x)$ of possible successor states at time $t + 1$ of x is obtained as follows:*

Decrease the residual work of stay–on customers in busy positions by one, converte the departure candidates into fresh jobs staying on their previous po-sitions and requesting for a further service of $K(m(i_l))$ time units, $l = 1, \ldots, k$. Then permute the stay–on customers on positions $\{1, \ldots, \ldots, n\} - \{i_1, \ldots, i_k\}$ according to any permutation resulting in state $y \in A(x)$. This happens with probability

$$c(n)d(x, y) \geq 0, \quad \sum_{y \in A(x)} d(x, y) = 1$$

(Note that the rearrangement of the stay–on customers may depend on the previous state of the node.)

From the description of the queueing discipline and the stochastic assumpti-ons put on the systems it follows that the time evolution of the system can be described by a homogeneous Markov chain $X = (X(t) : t \in I\!N)$ with state space S described above. Assuming the node to be doubly stochastic we shall prove the existence of a stationary measure, which is said to be of internal product form, for this Markov chain. The formal definition of being doubly stochastic will be given in the proof of the theorem.

Theorem 5.3 (Stationary measure) *The Markov chain $X = (X_t : t \in I\!N)$ describing a doubly stochastic node's evolution has a stationary measure $\hat{\pi} = (\hat{\pi}(x) : x \in S)$ on S given by*

$$\hat{\pi}([m(n), k(n); \ldots; m(1), k(1)]) = \frac{\prod_{h=0}^{n-1} b(h)}{\prod_{h=0}^{n} c(h)} \prod_{h=1}^{n} \frac{a(m(h))}{C(h)}, \quad (5.1)$$

$$[m(n), k(n); \ldots; m(1), k(1)] \in S.$$

Let $\mu = \sum_{m \in M} a(m)K(m)$ denote the mean service request of an average (typical) customer. An immediate consequence of theorem 5.3 is an insensitivity result similar to those obtained for non–FCFS nodes in BCMP networks and for Kelly's symmetric servers.

Corollary 5.4 (Steady state) *The stationary measure $\hat{\pi}$ can be normalized to a probability law $\pi = (\pi(x) : x \in S)$ if and only if $A = \sum_{n=0}^{\infty} \hat{\pi}(n) < \infty$ holds, where $\hat{\pi}(0) = c(0)^{-1}$ and for $n > 0$*

$$\hat{\pi}(n) = \left(\frac{\prod_{h=0}^{n-1} b(h)}{\prod_{h=0}^{n} c(h)} \right) \cdot (\mu - 1)^{n-C(n)} \cdot \mu^{C(n)} \cdot \prod_{i=1}^{n} C(i)^{-1}.$$

If the node is doubly stochastic and $A < \infty$ holds then the steady state π is given by

$$\pi([m(n), k(n); \ldots; m(1), k(1)]) = \frac{\prod_{h=0}^{n-1} b(h)}{\prod_{h=0}^{n} c(h)} \prod_{h=1}^{n} \frac{a(m(h))}{C(h)} A^{-1}, \quad (5.2)$$

$$[m(n), k(n); \ldots; m(1), k(1)] \in S.$$

The equilibrium queue length distribution π on $I\!N$ is

$$\pi(n) = \left(\frac{\prod_{h=0}^{n-1} b(h)}{\prod_{h=0}^{n} c(h)} \right) \cdot (\mu - 1)^{n-C(n)} \cdot \mu^{C(n)} \cdot \prod_{i=1}^{n} C(i)^{-1} \cdot A^{-1}. \quad (5.3)$$

The equilibrium queue length distribution is insensitive under perturbations of the service time distributions as long as their mean is fixed.

Proof of theorem 5.3: We follow the lines of the proof for the case of state independent arrival rate $b(n) = b$ in section 2 of [117] and use a partial balance technique which splits the global balance equations

$$\hat{\pi}(x) = \sum_{y \in S} \hat{\pi}(y) P(y, x), \quad x \in S, \quad (5.4)$$

into partial equations, where $(P(y, x) : y, x \in S)$ is the transition probability of X. If these partial equations share a common solution $\hat{\pi}$ they sum up to (5.4). The first equation of (5.4) for $x = e$ reads

$$\frac{1}{c(0)} = \frac{1}{c(0)} c(0) + \sum_{m \in M} \left(\frac{b(0)}{c(0)c(1)} a(m) \right) c(1). \quad (5.5)$$

Now let $x = [m(n), k(n); \ldots; m(1), k(1)]$ and assume that x shows no fresh jobs, i.e., $k(i) < K(m(i))$ for all positions $i = 1, \ldots, n$. The sum in the righthand side of (5.4) is split into two sums. One type of predecessor states $y \in A^0(x)$, are those states which are obtained from x by arbitrarily rearranging the customers on positions $1, \ldots, n$ and increasing the residual work by one for those customers staying now on the busy positions. Then the right hand side of (5.4) is

$$\sum_{y=[m'(n),k'(n);\ldots;m'(1),k'(1)]\in A^0(x)} \left(\frac{\prod_{h=0}^{n-1} b(h)}{\prod_{h=0}^{n} c(h)} \prod_{h=1}^{n} \frac{a(m(h))}{C(h)} \right) c(n)d(y,x), \quad (5.6)$$

where $d(y,x)$ is defined in definition 5.2, (A). If

$$\sum_{y\in A^0(x)} d(y,x) = 1, \quad (5.7)$$

then (5.6) reduces to

$$c(n)\hat{\pi}(x). \quad (5.8)$$

The second type of predecessor states $y \in A_{ri}^0(x)$, of x are those states which are obtained from x by inserting a fresh customer of type $r \in M$ in position i, $i \in \{1, \ldots, C(n+1)\}$, according to the shift protocol and then arbitrarily rearranging the customers on positions $1, \ldots, i-1, i+1, \ldots, n$ and increasing the residual work by one for those customers staying now on the busy positions. Then the right hand side of (5.4) is

$$\sum_{y=[m'(n),k'(n);\ldots;r,K(r);\ldots;m'(1),k'(1)]\in A_{ri}^0(x)} \left(\frac{\prod_{h=0}^{n} b(h)}{\prod_{h=0}^{n+1} c(h)} \right) \quad (5.9)$$
$$\cdot \left(\prod_{h=1}^{n} \frac{a(m(h))}{C(h)} \frac{a(r)}{C(n+1)} \right) c(n+1)d^-(y,x),$$

where $d^-(y,x)$ is defined in definition 5.2, (C). If

$$\sum_{y\in A_{ri}^0(x)} d^-(y,x) = 1, \quad (5.10)$$

then (5.9) reduces to

$$c(n)\hat{\pi}(x)\frac{b(n)a(r)}{C(n+1)}. \quad (5.11)$$

Summing (5.1) over all positions $i = 1, \ldots, C(n+1)$, and all types $r \in M$ yields the left hand side of (5.4).

Next we consider a generic state $x = [m(n), k(n); \ldots; m(1), k(1)]$ with exactly one fresh job which is of type r in position i, i.e., $m(i) = r$ and $k(i) = K(m(i))$, and for all other positions $j \neq i$ we have $k(j) < K(m(j))$.

The sum in the righthand side of (5.4) is split into two sums. One type of predecessor states $y \in A^0(x)$, are those states which are obtained from x by deleting the fresh customer and arbitrarily rearranging the other customers on positions $1, \ldots, n-1$ and increasing the residual work by one for those customers staying now on the busy positions. Then the right hand side of (5.4) is

$$
\sum_{y=[m'(n-1),k'(n-1);\ldots;m'(1),k'(1)]\in A^0(x)} \left(\frac{\prod_{h=0}^{n-2} b(h)}{\prod_{h=0}^{n-1} c(h)} \prod_{\substack{h=1 \\ h\neq i}}^{n} a(m(h)) \right)
$$
$$
\cdot \left(\prod_{h=1}^{n-1} \frac{1}{C(h)} \right) b(n-1)a(r) \frac{1}{C(n)} d^+(y,x), \tag{5.12}
$$

where $d^+(y,x)$ is defined in definition 5.2, (B). If

$$
\sum_{y\in A^0(x)} d^+(y,x) = 1, \tag{5.13}
$$

then (5.12) reduces to

$$
c(n)\hat{\pi}(x). \tag{5.14}
$$

The second type of predecessor states $y \in A^0_{r'_i}(x)$, of x are those states which are obtained from x by deleting the fresh customer of type r on position i and substituting him by a departure candidate of type $r' \in M$ and then arbitrarily rearranging the other customers on positions $1, \ldots, i-1, i+1, \ldots, n$, increasing the residual work by one for those customers staying now on the busy positions. Then the right hand side of (5.4) is

$$
\sum_{y=[m'(n),k'(n);\ldots;r',K(r');\ldots;m'(1),k'(1)]\in A^0_{r'_i}(x)} \left(\frac{\prod_{h=0}^{n-1} b(h)}{\prod_{h=0}^{n} c(h)} \prod_{\substack{h=1 \\ h\neq i}}^{n} a(m'(h))a(r') \right)
$$
$$
\cdot \left(\prod_{h=1}^{n} \frac{1}{C(h)} \right) b(n)a(r)d^+(y,x), \tag{5.15}
$$

where $d^+(y,x)$ is defined in definition 5.2, (D). If

$$
\sum_{y\in A^0_{r'_i}(x)} d^+(y,x) = 1, \tag{5.16}
$$

then (5.15) reduces to

$$
b(n)a(r')\hat{\pi}(x). \tag{5.17}
$$

Summing (5.17) over all types $r \in M$ and adding (5.14) yields the left hand side of (5.4) for this case.

The only remaining states are of the form $x = [m(n), k(n); \ldots; m(1), k(1)]$ with exactly $k \geq 2$ fresh customers in positions $i_1, \ldots, i_k$, being of type $m(i_l)$ in position i_l with $k(i_l) = K(m(i_l))$, and for all other positions $j \neq i_1, \ldots, i_k$ we have $k(j) < K(m(j))$.

The sum in the righthand side of (5.4) is split into two sums. One type of predecessor states $y \in A^0(x)$, are those states which are obtained from x by fixing the fresh customers on their positions with maximum residual work and arbitrarily rearranging the other customers on positions $\{1, \ldots, n\} - \{i_1, \ldots, i_k\}$, increasing the residual work by one for those customers staying now on the busy positions. Then the right hand side of (5.4) is

$$\sum_{y=[m'(n),k'(n);\ldots;m'(1),k'(1)]\in A^0(x)} \left(\frac{\prod_{h=0}^{n-1} b(h)}{\prod_{h=0}^{n} c(h)} \prod_{h=1}^{n} \frac{a(m(h))}{C(h)} \right) c(n)d(y,x), \quad (5.18)$$

where $d(y,x)$ is defined in definition 5.2, (E). If

$$\sum_{y\in A^0(x)} d(y,x) = 1, \quad (5.19)$$

then (5.18) reduces to

$$c(n)\hat{\pi}(x). \quad (5.20)$$

The second type of predecessor states $y \in A^0_{r'i_l}(x)$, of x are those states which are obtained from x by deleting the fresh customer of type $m(i_l)$ on position i_l and substituting him by a departure candidate of type $r' \in M$, fixing the residual fresh customers on their positions with maximal residual work and then arbitrarily rearranging the other customers on positions $\{1, \ldots, n\} - \{i_1, \ldots, i_k\}$, increasing the residual work by one for customers staying now on the busy positions. Then the right hand side of (5.4) is

$$\sum_{y=[m'(n),k'(n);\ldots;r',K(r');\ldots;m'(1),k'(1)]\in A^0_{r'i_l}(x)} \left(\frac{\prod_{h=0}^{n-1} b(h)}{\prod_{h=0}^{n} c(h)} \right) \quad (5.21)$$

$$\cdot \left(\prod_{\substack{h=1 \\ h\neq i_l}}^{n} a(m'(h))a(r') \right) \left(\prod_{h=1}^{n} \frac{1}{C(h)} \right) b(n)a(m(i_l))\frac{1}{k}d^+(y,x),$$

where $d^+(y,x)$ is defined in definition 5.2, (D). If

$$\sum_{y\in A^0_{r'i_l}(x)} d^+(y,x) = 1, \quad (5.22)$$

then (5.21) reduces to

$$b(n)a(r')\frac{1}{k}\hat{\pi}(x). \quad (5.23)$$

Summing (5.23) over all positions $i = 1, \ldots, k$ and all types $r' \in M$ and adding (5.20) yields the left hand side of (5.4) for this case. $\odot$

Definition 5.5 (Doubly stochastic discipline; Part III) *A queueing disciplin according to definitions 5.1 and 5.2 is doubly stochastic if it fullfills additionally (5.7), (5.10), (5.13), (5.16), (5.19), and (5.22).*

Example 5.6 (Doubly stochastic disciplines) *The class of* doubly stochastic *comprises especially nodes with the following queueing disciplines:* Last–Come–First–Served (preemptive resume) , infinite servers, random service allocation, round–robin with a preemptive modification for new arrivals *queueing regimes.*
First–Come–First–Served *is not included because immediate service must be guaranteed for a* doubly stochastic *node. For details see [117].*

The doubly stochastic servers can be considered to apply a permutation discipline to the customers in the node. This permutation should in general not being considered as a rule for physically redistributing stay–on customers at every time epoch. The more common interpretation is that the service capacity of the node is redistributed to the customers present. E.g., applying a suitable permutation rule would guarantee fairness of service for all customers entering the node. This was discussed e.g. for the case of *processor sharing* by Kleinrock [82], pp. 166–172. Processor sharing in its classical definition for the continuous time framework is fair without applying a permutation rule because of the homogeneous distribution of service capacity among the customers in system.
Permutation rules for general symmetric servers in isolation were introduced by Yashkov [143]. Moreover, these permutation rules open the possibility to control the service process nearly continuously.

5.2 Description of the Geometrical Server

In this section the analogue of the exponential FCFS nodes in the BCMP networks or, more general, of Kelly's general exponential servers is investigated. These nodes are fed by a state dependent arrival stream of customers which are of different types $m \in M$, M being a countable set of types.

The node consists of an unlimited sequence of service and waiting positions $1,2,3, \ldots$ which is controlled according to the *shift–protocol*. Whenever there are n customers present, $n \geq 1$, they will occupy positions $1, \ldots, n$. If the customer in position $i \in \{1, \ldots, n\}$ departs from the node then the gap is closed by shifting the customers previously on positions $i + 1, \ldots, n$ one step down into positions $i, \ldots, n - 1$, leaving their internal order invariant. Likewise if an additional customer is inserted into position $i \in \{1, \ldots, n\}$, the customers previously on positions $i, \ldots, n$ are shifted one step up into positions $i+1, \ldots, n+1$, leaving their internal order invariant.

Having $n > 0$ customers present at the node service is provided to those customers staying on positions $1, \ldots, C(n)$, where $C(n) > 0$ is a node specific service parameter. We call positions $1, \ldots, C(n)$ *busy*, while positions $C(n) + 1, \ldots, n$ are said to be *idle*. $(C(0) := 0.)$

Customers entering the node request for an amount of service time which will be provided to them according to the service disciplin (to be described below) of the node.

If at time $t \in I\!N$ there are n customers present then there will be no arrival with probability $c(n) = (1 - b(n))$ – or there is exactly one arrival of type m with probability $b(n) \cdot a(m) > 0$, $m \in M$, such that $\sum_{m \in M} a(m) = 1$ holds.

We assume the arrival decisions for a time slot to depend on the history of the system only through the total number of customers in system at the beginning of that time slot, and the type decisions to be independent of the history of the system. We assume late arrivals and for multiple events we assume the D/A rule (departure before arrival). The state of the system is recorded at times $t \in I\!N$, just after posible departures and arrivals have happened.

The amount of service a customer requests for is geometrically distributed with a node characteristic parameter $p \in (0,1)$ for all types of customers. The sampling of service times is independent and independent from the arrival process.

Following from this description a suitable state space S for a Markovian description of the node's evolution over time $I\!N$ consists of elements x as follows:

$x := e$ for the empty node, and sequences

$x := [m(n); \ldots ; m(1)], \quad n \geq 1$

where n is the number of customers present at the node (*queue length of the node*), $m(i)$ is the type of the customer on position i. (Due to the memoryless property of the geometrical distribution we need no information about residual work or age of service time.)

Definition 5.7 (Geometrical discipline)

(1) A customer present at time t in a busy position $i \in \{1, \ldots, C(n)\}$ obtains exactly one unit of service time until time $t+1$. With probability p his service ends at the end of $[t, t+1)$, and he departs from the node. (But notice the restriction on the departure rules given below in (4).) With probability $q = 1 - p$ he requests for at least one more service quantum.

(2) An arriving customer observing n other customers already present enters position $n+1, n \geq 0$. (But notice the restriction on the arrival rules given below in (4)) Applying the shift–protocol then leads to a FCFS–based service.

(3) If more than one customer complete their service at the same time instant they are not allowed to depart jointly. They have to stay at their present position for obtaining just another service (retrial of transmission).

(4) If an arrival occurs and at least one services expires jointly at the same time instant then the departure candidates stay for another service on their positions and the arrival candidate is rejected and lost.

Therefore either one single external arrival or at most one service completion with a subsequent departure from the node can be observed.

From the description of the queueing discipline and the stochastic assumptions put on the systems it follows that the time evolution of the system can be described by a homogeneous Markov chain $X = (X(t) : t \in I\!N)$ with state space S described above. To prove the existence of a stationary measure is done by direct verification of (5.24) as a solution of the steady state equation. This solution is said to be of internal product form again.

Theorem 5.8 (Stationary measure) *The Markov chain $X = (X_t : t \in I\!N)$ describing the evolution of a geometrical node has a stationary measure $\hat{\pi} = (\hat{\pi}(x) : x \in S)$ on S given by*

$$\hat{\pi}([m(n); \ldots; m(1)]) \qquad [m(n); \ldots; m(1)] \in S, \qquad (5.24)$$

$$= \left(\frac{\prod_{h=0}^{n-1} b(h)}{\prod_{h=0}^{n} c(h)} \right) \prod_{h=1}^{n} \frac{a(m(h))}{C(h)} \left(\frac{q}{p} \right)^{n} \left(\frac{1}{q} \right)^{C(n)}.$$

An immediate consequence of theorem 5.8 is an ergodicity criterion for geometrical nodes.

Corollary 5.9 (Steady state) *The stationary measure $\hat{\pi}$ can be normalized to a probability law $\pi = (\pi(x) : x \in S)$ if and only if $A = \sum_{n=0}^{\infty} \hat{\pi}(n) < \infty$ holds, where $\hat{\pi}_j(0) = c(0)^{-1}$ and for $n > 0$*

$$\hat{\pi}(n) = \left(\frac{\prod_{h=0}^{n-1} b(h)}{\prod_{h=0}^{n} c(h)} \right)^{n} \prod_{i=1}^{n} C(i)^{-1} \left(\frac{q}{p} \right)^{n} \left(\frac{1}{q} \right)^{C(n)}.$$

If $A < \infty$ holds then the steady state π is given by

$$\pi([m(n); \ldots; m(1)]) = \frac{\prod_{h=0}^{n-1} b(h)}{\prod_{h=0}^{n} c(h)} \prod_{h=1}^{n} \frac{a(m(h))}{C(h)} \left(\frac{q}{p} \right)^{n} \left(\frac{1}{q} \right)^{C(n)} A^{-1}$$

$$[m(n); \ldots; m(1)] \in S. \qquad (5.25)$$

The equilibrium queue length distribution π on $I\!N$ is

$$\pi(n) = \left(\frac{\prod_{h=0}^{n-1} b(h)}{\prod_{h=0}^{n} c(h)} \right)^{n} \prod_{i=1}^{n} C(i)^{-1} \left(\frac{q}{p} \right)^{n} \left(\frac{1}{q} \right)^{C(n)} A^{-1}, \qquad n \in I\!N. \qquad (5.26)$$

Remark 5.10 (Slotted Aloha–type protocol) *The restriction concerning the departure and arrival rule for the geometrical node can be interpreted as describing transmission stations in a slotted Aloha–type communication system, [82], section 5.11, [142], section 6.2. If there are $C(n)$ sources of traffic (stations), and the end of a service in position $i \in \{1, \ldots, C(n)\}$ indicates that a message has to be transmitted over the shared medium, then this is possible if and only if exactly one service ends. If more than one service ends, and more than one message is tried to be send at the same time instant, all those transmission trials are not successful and have to be repeated. A common regime to resolve the conflicts is that the sources retry at random to repeat sending. This is just what is going on in a geometrical server according to (3) in definition 5.7.*
Due to the memoryless property of the geometrical service time distribution the blocking mechanism according to repititive service is equivalent to what is known as communication blocking, [105], p.455.

5.3   Open Networks of Doubly Stochastic and Geometrical Nodes

Performance prediction for complex networks in computer and communication systems is usually done either by simulation or by using exact or approximative analytical methods. The usefulness of analytical methods strongly relies on having simply structured models at hand. Or, saying it the other way round: In the modeling process we have to impose structural assumptions on the systems which in fact are approximations of the more complicated direct description. The networks under consideration in this section show several such approximations:

First: They are constructed along the lines of the the celebrated BCMP networks [4] and Kelly's networks [75]. These networks are now widely accepted as a versatile class of queueing networks, which simulate sufficiently detailed the behaviour of many complex systems. The latter class is a generalisation of the former one, because it removes the restriction on allowing only *processor sharing, Last–Come–First–Served(preemptiveresume), infinite server , and exponential First–Come–First–Served* regimes for the nodes by introducing the so–called *general exponential* and *symmetric servers*.

Second: We substitute the symmetric and exponential servers by doubly stochastic and geometrical servers, where multiple events are controlled by the rules described insections 5.1, 5.2.

Third: We shall impose a similar global regulation regime for multiple events originating from the synchronized behaviour of the different nodes.

Due to these modeling assumptions not only the steady state probabilities of the networks can be computed explicitly, but main performance quantities are directly accessible as well. An appealing property is that the first order mean values of the relevant performance measures are insensitive: They remain invariant under variation of shape of the customers' service time distribution at doubly stochastic servers as long as the mean service time remains invariant.

The network of queues in this section consists of J nodes and is fed by a Bernoulli arrival stream of customers which are of different types $m \in M$, M being a countable set of types (not containing natural numbers for making the later state description unambiguous).

At any time $t \in I\!N$ there is either no arrival, with probability $c = 1 - b$, or there is exactly one arrival, being of type m with probability $b \cdot a(m) > 0$, $m \in M$, such that $\sum_{m \in M} a(m) = 1$ holds. We assume the successive arrival and type decisions to be an independent sequence being independent of the previous history of the network.

The type m of an arriving customers uniquely specifies the route of this customer through the network: $W(m) = (W(m,1), W(m,2), \ldots, W(m, S(m)))$, where node $W(m,i)$ is the ith stage of m on his itinerary, and $1 \leq S(m) < \infty$ is the length of his route. We assume for simplicity of presentation that $W(m,i) \neq W(m, i+1), 1 \leq i < S(m)$. (For a short description of how to remove this restriction see [33].)

The nodes of the network are of two different types:

Nodes $1, \ldots, J', 0 \leq J' \leq J$, are *doubly stochastic* (see section 5.1), and resemble Kelly's *symmetric nodes* in continuous time queueing systems (see [75], Chapter 3.1 and 3.3). Nodes $J'+1, \ldots, J$ are geometrical nodes (see section 5.2), which are the counterparts of exponential multiserver and general exponential servers, respectively.

If $n_j > 0$ customers are present at node j service is provided to those customers staying on positions $1, \ldots, C(j, n_j)$, where $C(j, n_j) > 0$ is a node specific service parameter. We then call positions $1, \ldots, C(j, n_j)$ *busy*, while positions $C(j, n_j) + 1, \ldots, n$ are said to be *idle*. (We set $C(j, 0) := 0, 1 \leq j \leq J$.)

Customers entering node $j \in \{1, \ldots, J\}$ request at this node for an amount of service time which will be provided to them according to the service discipline of that node.

The amount of service time a customer requests for at a geometrical node j is node specific geometically distributed on $I\!N_+ = \{1, 2, \ldots\}$. A customer requests with probability $p_j(1 - p_j)^{k-1}$, $p_j \in (0, 1]$, for exactly k time units of service at node j, $k \in I\!N_+$. These geometrical service times are drawn independently and independent of anything else in the history of the network.

The amount of service a customer requests for at a doubly stochastic node j depends deterministically on the customer's type and the stage number he has reached on his passage through the network. If a customer of type $m, m \in M$, enters stage $s, 1 \leq s \leq S(m)$, of his route, which is according to the definition $W(m, s) = j$, he will request for an amount of $K(m, s)$ units of service time, $K(m, s) \in I\!N_+$.

We assume that for every doubly stochastic node there exists at least one customer who requests at this node for more than one unit of service time.

The requirement of having deterministic service times at doubly stochastic nodes is not a restriction but widens considerably the possibility of modeling the probabilistic behaviour of customers at doubly stochastic nodes. The key is the introduction of different customer types and of the type and stage dependent behaviour: The random decision for the successive service requests at doubly stochastic nodes on a customer's route is done by selecting the customer's type when entering the network.

To be more specific: Let us assume that we have a set of customers with different (physical) customer types requesting for service according to general (type– and stage–dependent) distributions at the doubly stochastic nodes of their route. We can discriminate between different sampled sequences of requests for a specific customer type by introducing ficticious customer types. Each ficticious customer type carries information about the physical type of that customer, his routing, and his exact successive amounts of requested service at his successive stages on his itinerary. This concept even allows for using stochastically dependent service requests at the successive doubly stochastic nodes on a customer's itinerary.

It should be noted that a similar procedure for determining the service requests is possible in a network with random routing. Then different routes are distinguished by using types which include fixed route descriptions:
These types are determined for any customer in advance when entering the network. Their routes (and their types) are sampled according to the Markov chain determined by the routing matrix.

Following from the above description the construction of a suitable state space for a Markovian description of the network's evolution over time should include at least the following information – from the description of the service disciplines below it will follow that this information does already suffice:

The local state space S_j for a doubly stochastic node $j, j \in \{1, \ldots, J'\}$ consists of elements x_j as follows:

$x_j := e_j$ for the empty node, and sequences
$$x_j := [m(j, n_j), s(j, n_j), k(j, n_j); \ldots; m(j, 1), s(j, 1), k(j, 1)], \quad n_j \geq 1$$
where n_j is the number of customers present at node j (*queue length at node j*), $m(j, i)$ is the type of the customer on position i, $s(j, i)$ is the stage number of this customer on his actual route, and $k(j, i)$ is his residual request for service time (his *residual work*), $1 \leq i \leq n_j$.

The local state space S_j for a geometrical node $j, j \in \{J' + 1, \ldots, J\}$ consists of elements x_j as follows:

$x_j := e_j$ for the empty node, and sequences
$$x_j := [m(j, n_j), s(j, n_j); \ldots; m(j, 1), s(j, 1),], \quad n_j \geq 1$$
where n_j is the number of customers present at node j (*queue length at node j*), $m(j, i)$ is the type of the customer on position i, and $s(j, i)$ is the stage number of this customer on his actual route, $1 \leq i \leq n_j$. (Here the residual workloads need not be included in the state description due to the memoryless property of the geometrical service time distributions at the nodes.)

Global states of the network are composed of these local states. The state space of the network is therefore $S := S_1 \times S_2 \times \cdots \times S_J$ or a subset thereof.

The service disciplines at the nodes are either doubly stochastic as described in section 5.1 or geometrical as described in section 5.2. It turns out that in case of a network with general toplogy we have to define some further rules which regulate the network's behaviour at instances of simultaneous events. This is described in the following four–step procedure.

Definition 5.11 (Queueing disciplines in the network) *Throughout we assume $x = (x_1, \ldots, x_J) \in S$ to be a generic state of the network as described above and consider the possible state changes out of x:*

(1) A customer present at time t in a busy position $i \in \{1, \ldots, C(j, n_j)\}$ of node j obtains until time $t + 1$ exactly one unit of service time.
If node j is geometric, this customers will leave node j thereafter (at time $(t + 1)-$) with probability p_j and will stay there for at least one further time unit with probability $1 - p_j$.
If node j is doubly stochastic , assume that the customers show at time t a residual work of $k(j, i)$.

If $k(j,i) > 1$ then at time $(t+1)-$ this residual workload is diminished to $k(j,i) - 1$.

If $k(j,i) = 1$ then this customer departs from node j at time $(t+1)-$. (Note the restrictions as consequences of simultaneous events prescribed further for the departure candidates in (3) below.)

(2) A customer of type m on stage s of his route enters node $j := W(m,s)$, seeing just before his entrance (which will happen between some times $t-$ and t) that node to be in local state x_j.

For j being a geometrical node: if $x_j = e_j$, then the lokal state of node j changes to $[m,s]$; else if $x_j := [m(j,n_j), s(j,n_j); \ldots; m(j,1), s(j,1)]$, then the local state changes to

$$x_j := [m, s; m(j,n_j), s(j,n_j); \ldots; m(j,1), s(j,1)].$$

I.e., for geometrical nodes we have a FCFS–regime.

For j being a doubly stochastic node: if $x_j = e_j$, then the lokal state of node j changes to $[m, s, K(m,s)]$;
else if $x_j := [m(j,n_j), s(j,n_j), k(j,n_j); \ldots; m(j,1), s(j,1), k(j,1)]$, then with probability $1/C(j,n_j+1)$ the local state changes to

$$\begin{aligned}
x_j := \quad & [m(j,n_j), s(j,n_j), k(j,n_j); \ldots; m(j,i+1), s(j,i+1), k(j,i+1); \\
& m, s, K(m,s); m(j,i-1), s(j,i-1), k(j,i-1); \\
& \ldots; m(j,1), s(j,1), k(j,1)],
\end{aligned}$$

for some $i \in \{1, \ldots, C(j, n_j+1)\}$. I.e., for doubly stochastic nodes a new arrival is inserted with probability 1 into a busy position and therefore has immediate access to service capacity. His position is randomly selected among the busy positions.

(3) Due to the discrete time scale at any time point there may occur simultaneous or multiple events composed of several arrival and departure events. Our rule to handle these multiple events resembles what is as repititive service/rejection blocking in multiple access transmission systems with limited buffer capacity. The details:

If a multiple event occurs where several customers simultaneously wish to enter and/or leave nodes of the network, then none of the requested transitions is allowed:

Customers on arrival from the outside source are lost (restarted); customers wishing to depart from the network due to a service completion at node $W(\cdot, S(\cdot))$, or wishing to enter the next stage of their route due to service completion on the present stage have to stay at their present node on their present position for obtaining just another service (retrial of transmission) .

This request for retrial service time is distributed acccording to the node specific geometrical distribution, if the node is geometrical. Otherwise, if the node is doubly stochastic, then the additional service request is deterministically selected and identical to the previous service there. (The additional service at the node is therefore not counted as an additional stage for the customer's passage.)

Therefore at most one single external arrival or one service completion with a subsequent departure from some node can be observed. It follows that no access conflicts with respect to the handling of multiple events will occur.

(4) Assume that according to (1) or (2) exactly one arrival or one service completion or according to (3) a multiple event appeared and is handled. Then all the stay–on customers at doubly stochastic nodes, i.e., those customers staying before on idle positions and customers on busy positions whose service did not expire, are permuted on their positions (nodewise) according to some probability law. This law may depend on what has happend in (1),(2),(3) and on the global state of the network.

The only restriction is: Given an event as described in (1),(2),(3), then the transition matrix which governs the allowed state changes due to the permutations is doubly stochastic in a similar way as defined in definition 5.5 and the proof of theorem 5.3. (For details see [33].)

The continuous time counterparts of the networks with doubly stochastic and geometrical nodes were defined in [75] and [4] without the permutation rules for redistributing customers. This additional feature was invented by Yaskov [143] for single node systems, and for general networks in [25], for a description in a more general network environment see [30].

As can be seen from the definition, the permutation rules introduce rather general control schemes for discrete time networks: Redistributing customers among the busy positions or laying customers dormant on idle positions for some time, generates state dependent schemes for providing service to the customers. According to the general decision rules for applying the permutations these rules may take into consideration the states of e.g. neighboured nodes for deciding which customers to serve.

The restriction for movements in the network consisting of either a single arrival or a single departure from exactly one node according to the blocking protocol (repititive service/rejection blocking) resembles the departure protocol applied in [100] for batch service disciplines in discrete time networks. At every time epoch at most one node is selected to release a batch of customers being served to be distributed over the network or partially to leave the network. As Miyazawa puts it, this model is motivated not only by the fact that it is important for discrete time queueing networks, but also by its tractability for analysis. This statement applies here and to the following sections of the chapter as well.

5.4   Stationary State for the Open Network with Doubly Stochastic and Geometrical Nodes

A detailed description of the transition laws which are sketched in section 5.3 is given in [33] in connection with solving the equilibrium equation of the Markov chain $X = (X_t : t \in I\!N)$ which describes the time evolution of the network.

The state space $S = S_1 \times \cdots \times S_J$ is not minimal. E.g., a customer of type $m \in M$ present at a doubly stochastic node $W(m, s) = i \leq J'$, can show a (maximal) residual work of $K(m, s)$ time units if and only if he is on a busy position of j.

We assume henceforth that S and $S_j, 1 \leq j \leq J$ are restricted to those states showing feasible workloads for the customers present there. We define local measures for the nodes as follows:

If j is a doubly stochastic node then $\hat{\pi}_j = (\hat{\pi}_j(x_j) : x_j \in S_j)$ is given by

$$\hat{\pi}_j(e_j) = 1, \qquad and$$

$$\hat{\pi}_j([m(j, n_j), s(j, n_j), k(j, n_j); \ldots; m(j, 1), s(j, 1), k(j, 1)]) \qquad (5.27)$$

$$= \left(\frac{b}{c}\right)^{n_j} \prod_{i=1}^{n_j} \frac{a(m(j, i))}{C(j, i)}$$

If j is a geometrical node then $\hat{\pi}_j = (\hat{\pi}_j(x_j) : x_j \in S_j)$ is given by

$$\hat{\pi}_j(e_j) = 1, \qquad and$$

$$\hat{\pi}_j([m(j, n_j), s(j, n_j); \ldots; m(j, 1), s(j, 1)]) \qquad (5.28)$$

$$= \left(\frac{b}{c}\right)^{n_j} \left(\frac{q_j}{p_j}\right)^{n_j} \left(\frac{1}{q_j}\right)^{C(j, n_j)} \prod_{i=1}^{n_j} \frac{a(m(j, i))}{C(j, i)}$$

Theorem 5.12 (Stationary measure) *The Markov chain $X = (X_t : t \in I\!N)$ describing the network's evolution has a stationary measure $\hat{\pi} = (\hat{\pi}(x) : x \in S)$ on S given by*

$$\hat{\pi}(x) = \prod_{j=1}^{J} \hat{\pi}_j(x_j) \quad , \qquad if \quad x = (x_1, \ldots, x_J) \in S. \qquad (5.29)$$

The proof of the theorem is along the lines of the proof of theorem 5.3: Identifying suitable partial balance equations which are solved by (5.29) individually and which sum up to a solution of the global balance equations.

Now let

$$b_j = \sum_{(m,s):W(m,s)=j} b \cdot a(m), \qquad 1 \leq j \leq J,$$

denote the total arrival rate at node j, and μ_j the mean service request of a typical customer at node j: For a geometrical node j $\mu_j = p_j^{-1}$ and for a doubly stochastic node j $\mu_j = \sum_{(m,s):W(m,s)=j} ba(m)b_j^{-1}K(m, s)$.

Corollary 5.13 (Steady state) *The stationary measure $\hat{\pi}$ can be normalized to a probability law $\pi = (\pi(x) : x \in S)$ if and only if $\sum_{n=0}^{\infty} \hat{\pi}_j(n) < \infty$ holds for all j, where $\hat{\pi}_j(0) = 1$ and for $n > 0$*

$$\hat{\pi}_j(n) = \left(\frac{b_j(\mu_j - 1)}{c}\right)^n \cdot \left(\frac{\mu_j}{\mu_j - 1}\right)^{C(j, n)} \cdot \prod_{i=1}^{n} C(j, i)^{-1} \quad .$$

5.5   Performance Measures for Open Networks

We assume henceforth that normalization of $\hat{\pi}$ is possible. This is the case if and only if the local measures $\hat{\pi}_j : j = 1, \ldots, J$ are normalizable as well, obtaining local probability laws $\pi_j : j = 1, \ldots, J$. Starting X with distribution π then yields X to be stationary with $Pr\{X_t = x\} = \pi(x), \quad x \in S$. We refer to this situation in the following as X to be in *steady–state* and derive the main *steady–state performance measures* of the network described by X. The proofs can be found in [33].

Corollary 5.14 (Steady state) *For X being in steady–state let*
$A_j = \sum_{n=0}^{\infty} \hat{\pi}_j(n), j = 1, \ldots, J$ *and* $A = \prod_{j=1}^{J} A_j$. *Then*
$\quad \pi_j(x_j) = \hat{\pi}_j(x_j) \cdot A_j^{-1}$ *is the steady-state probability of finding node j in local state $x_j \in S_j$ and*
$\quad \pi_j(n) = \hat{\pi}_j(n) \cdot A_j^{-1}, n \in I\!N$ *is the steady-state queue length probability at node $j, j = 1, \ldots, J$. We further have*
$\quad Pr\{X_t = x\} = \pi(x) = \hat{\pi}(x) \cdot A^{-1}, \quad x \in S.$

Remark 5.15 (Insensitive queue length distributions) *The probability to see in steady state a queue length vector $(n_1, \ldots, n_J)$ is*

$$\prod_{j=1}^{J} \pi_j(n_j)$$

obtained from Corollary 5.14. From the explicit form in Corollary 5.13 it follows that this probability is insensitiv: *It depends on the requested service times only through their means. So the stationary queue length distribution remains invariant if the service time distribution of a physical customer for doubly stochastic nodes is changed as long as the mean remains invariant.*
This remark applies for random service times as well. (See the remarks in section 5.3, page 87.)
Related insensitivity results for single node systems in discrete time are proved in [43].

Corollary 5.16 (Loss probability) *The steady–state loss probability for arrivals from the outside of the network is*

$$\pi_l = 1 - \prod_{j=1}^{J} A_j^{-1} \cdot B_j, \tag{5.30}$$

where

$$B_j = \sum_{n=0}^{\infty} \left(\prod_{i=1}^{n} C(j,i)^{-1} \right) \left(\frac{b_j(\mu_j - 1)}{c} \right)^n, \quad 1 \le j \le J. \tag{5.31}$$

Corollary 5.17 (Mean queue length and node delay) *Denote by*

$$\mu(m, s) = \begin{cases} K(m, s) & if \ W(m, s) \leq J' \quad (doubly \quad stochastic \quad node), \\ p_{W(m,s)}^{-1} & if \ W(m, s) > J' \qquad (geometrical \quad node), \end{cases}$$

the mean requested service time of a customer of type m on stage s of his route. The steady–state mean number of customers of type m in stage s of their route is

$$L(m, s) \qquad (5.32)$$
$$= \ b \cdot a(m) \cdot (\mathcal{A}(W(m, s))\mu(m, s) - \mathcal{B}(W(m, s))), \quad m \in M, \quad 1 \leq s \leq S(m),$$

where

$$\mathcal{A}(j) \ = \ A_j^{-1} \sum_{n=1}^{\infty} (\prod_{i=1}^{n} C(j, i)^{-1}) b_j^{-1} \left(\frac{b_j}{c}\right)^n$$
$$\cdot \mu_j^{C(j,n)-1} (\mu_j - 1)^{n-1-C(j,n)} (n\mu_j - C(j, n))$$

and

$$\mathcal{B}(j) \ = \ A_j^{-1} \sum_{n=1}^{\infty} (\prod_{i=1}^{n} C(j, i)^{-1}) b_j^{-1} \left(\frac{b_j}{c}\right)^n \qquad (5.33)$$
$$\cdot \mu_j^{C(j,n)} (\mu_j - 1)^{n-1-C(j,n)} (n - C(j, n)).$$

The mean sojourn time $T(m, s)$ (waiting time + service time) of a customer of type m on stage s of his route at node $W(m, s)$ is

$$T(m, s) \qquad (5.34)$$
$$= \ \mathcal{A}(W(m, s))\mu(m, s) - \mathcal{B}(W(m, s)), \quad m \in M, \quad 1 \leq s \leq S(m).$$

Corollary 5.18 (Mean response time) *The steady–state mean response time (passage time) $T(m)$ of a type m customer is*

$$T(m) = \sum_{s=1}^{S(m)} (\mathcal{A}(W(m, s))\mu(m, s) - \mathcal{B}(W(m, s))), \qquad (5.35)$$

where the response time is counted as zero if the job is lost. Conditioned on a customer being accepted by the network his response time is $((5.35) \times (1 - \pi_l)^{-1})$.

Remark 5.19 (Conditional response time) *A customer's mean sojourn time $T(m, s)$ (5.34) is an affin–linear function of his requested mean service time at the node $W(m, s)$. A customer's mean response time $T(m)$ is an affin–multilinear function of his successive mean service requests at the nodes of his path $W(m)$.*

Recall that the requested service times at doubly stochastic nodes are determi- nistic $K(m, s)$. As we pointed out in Section 5.3 this can be used as a modeling device for handling the distribution of complex, even correlated, service times by *chosing suitable fictitious customer types. Then $\mu_{W(m,s)} = K(m, s)$ in (5.33) and (5.34) for doubly stochastic nodes $W(m, s)$ is a conditioning event. It follows:* Conditional mean response times of a physical customer are affin–multilinear in his actual service time requests at the doubly stochastic nodes.

Parallel results for continuous time networks with symmetric nodes (including the special cases of Baskett, Chandy, Muntz, and Palacios [4], can be found in [22] and [3]. The proofs are much more elaborated then their discrete time counterparts presented here. A detailed discussion of this property for single node systems with respect to performance analysis and fair service regimes can be found in [82], pp.166–172.

Remark 5.20 *The brute force resolution of multiple event access conflicts described under (3) in Section 5.3 results in explicit performance measures for our discrete time networks. The impact of this conflict resolution scheme is neglectible e.g. with respect to individual loss probabilities if the mean service times of a typical customer at the nodes is sufficiently large. We then have in Corollary 5.16 that B_j in (5.31) is appoximately A_j and in (5.30) $\pi_l \simeq 0$.*

Similarly to the simplifications discussed in Remark 5.20 the main performance measures of the network simplify considerably under the

Assumption 5.21 (Small time slots) *The mean requested service time of a typical customer at the nodes is sufficiently large, which means $\mu_j \approx \mu_j - 1$,*

Corollary 5.22 *Assume that the assumption 5.21 holds.*
Then for $j = 1, \ldots, J$

$$\mathcal{A}(j) = \frac{1}{A_j b_j} \sum_{n=1}^{\infty} (\prod_{i=1}^{n} \frac{\mu_j}{C(j,i)}) \left(\frac{b_j}{c} \right)^n (n\mu_j - C(j,n)),$$

and

$$\mathcal{B}(j) = \frac{\mu_j}{A_j b_j} \sum_{n=1}^{\infty} (\prod_{i=1}^{n} \frac{\mu_j}{C(j,i)}) \left(\frac{b_j}{c} \right)^n (n - C(j,n)).$$

Insert this into (5.32), (5.34), and (5.35) to obtain the simplified performance approximations.

Remark 5.23 *The rules to handle multiple events can be interpreted as follows: We introduce a dependence between the nodes to guarantee that the arrival and departure structures and in general, the customer flows in the network, will be simple. The question arises whether this rule, described in (3) of definition 5.11, Section 5.3, is the only possible one, which provides us with explicit product form equilibria.*

The answer is "No". This can be seen from the detailed investigation of the isolated doubly stochastic node in [117], see section 5.1 above. There it is allowed that at most one departure may concurrently occur with a single arrival. The steady state then is of product form. Even more can be concluded from Theorem 3 in [117]: If we consider an open linear tandem of doubly stochastic nodes with capacity 1, then in steady state the queue lengths behave as if they are independent (see section 5.7).

Incorporating simultaneous jumps of many customers between different nodes into the more general networks with correlated service requests, global reshuffling regimes for stay–on customers, and general topologies considered here, leads to an enormous combinatorial complexity and is an open problem for further research. For a discussion of this problem see [61], Section 8. There a connection to continuous time networks is pointed out, which indicates how to obtain further discrete time systems with product form equilibrium.

5.6 Closed Networks of Doubly Stochastic and Geometrical Nodes

In this section we consider a closed network of geometrical and doubly stochastic nodes as described in section 5.1 and 5.2. These are networks with a structure as described in section 5.3 without external arrivals, but with a fixed set of K customers cycling in it. We assume the nodes of the networks to be numbered as in section 5.3, nodes 1 through J' are doubly stochastic, nodes $J' + 1$ through J are geometrical. The characteristics of the nodes are those of section 5.3. The finite routes of the customers are completely determined by their types:

$$W(m) = (W(m,1), W(m,2), \ldots, W(m, S(m))), \text{ where } W(m,i) \text{ is the } i\text{th}$$
stage of m on his itinerary, and $1 \leq S(m) < \infty$ is the length of this route. We assume again for simplicity of presentation that $W(m,i) \neq W(m, i+1), 1 \leq i < S(m)$.

After departing from a route the customer changes his type according to some Markovian rule:
A customer being of type m on leaving node $W(m, S(m))$ becomes a customer of type m' with probability $r(m, m') \geq 0$, $m, m' \in M$, where $R = (r(m, m') : m, m' \in M)$ is a stochastic matrix. We assume R to be irreducible on M and positive recurrent. This is the single chain case as discussed in section 3.1. The multichain framework (see section 3.2) yields results similar to those to be presented below.

We assume that for every doubly stochastic node there exists at least one customer who requests at this node for more than one unit of service time. A special case of this class of networks was investigated by [118], the general formulas of this subsection were obtained by Krüger [89].

We denote the state space of the network by $\tilde{S}(K, J)$ which is a subset of $S = S_1 \times \cdots \times S_J$, resembling the population size K and the network size J in the denotation. We use the same notations as in section 5.3 and assume minimality and irreducibility of the state space, i.e., any state in $\tilde{S}(K, J)$ appears with positive probability. The states and their interpretation are the same as in Section 5.3.

Theorem 5.24 (Stationary measure) *Let* $\lambda = (\lambda(m) : m \in M)$ *be the probability solution of the system of traffic equations*

$$\lambda = \lambda \cdot R.$$

Let the local measures for the nodes be defined as follows:
If j is a doubly stochastic node then $\hat{\pi}_j = (\hat{\pi}_j(x_j) : x_j \in S_j)$ is given by

$$\hat{\pi}_j(e_j) = 1, \qquad and$$

$$\hat{\pi}_j([m(j, n_j), s(j, n_j), k(j, n_j); \ldots ; m(j, 1), s(j, 1), k(j, 1)]) \qquad (5.36)$$

$$= \prod_{i=1}^{n_j} \frac{\lambda(m(j, i))}{C(j, i)}.$$

If j is a geometrical node then $\hat{\pi}_j = (\hat{\pi}_j(x_j) : x_j \in S_j)$ is given by

$$\hat{\pi}_j(e_j) = 1, \qquad and$$

$$\hat{\pi}_j([m(j, n_j), s(j, n_j); \ldots ; m(j, 1), s(j, 1)]) \qquad (5.37)$$

$$= \left(\frac{q_j}{p_j}\right)^{n_j} \left(\frac{1}{q_j}\right)^{C(j, n_j)} \prod_{i=1}^{n_j} \frac{\lambda(m(j, i))}{C(j, i)}.$$

(Note that the local measures do not reflect the restriction put on the sytem by the finite population size.)
Then the Markov chain $X = (X_t : t \in \mathbb{N})$ describing the network's evolution has a stationary measure $\hat{\pi} = (\hat{\pi}(x) : x \in \tilde{S}(K, J))$ on $\tilde{S}(K, J)$ given by

$$\hat{\pi}(x) = \prod_{j=1}^{J} \hat{\pi}_j(x_j) \quad , \qquad if \quad x = (x_1, \ldots, x_J) \in \tilde{S}(K, J). \qquad (5.38)$$

The total arrival rate at node j,

$$b_j = \sum_{(m,s):W(m,s)=j} \lambda(m), \qquad 1 \le j \le J \quad ,$$

and the mean service request μ_j of a typical customer at node j is: For a geometrical node j

$$\mu_j = p_j^{-1}$$

and for a doubly stochastic node j

$$\mu_j = \sum_{(m,s):W(m,s)=j} \lambda(m) b_j^{-1} K(m, s).$$

Because the set of customer types may be infinite the state space of X is in general not finite. So we need a criterion for the positive recurrence of the states of X in $\tilde{S}(K, J)$.

Corollary 5.25 (Steady state) *The stationary measure $\hat{\pi}$ can be normalized to a probability law $\pi = (\pi(x) : x \in \tilde{S}(K, J))$ if and only if for all $j, j = 1, \ldots, J$*

$$b_j < \infty \quad \text{and} \quad \mu_j < \infty$$

. *A sufficient condition for positive recurrence is therefore the existence of constants such that*

$S(m) < \bar{S} < \infty, \quad m \in M$ *and*
$K(m, s) < \bar{K} < \infty, \quad (m, s) \quad \text{with} \quad W(m, s) \leq J'.$
The norming constant then is

$$G(K, J) = \sum_{x \in \tilde{S}(K, J)} \hat{\pi}(x).$$

Computing the norming constant can be performed by an algorithm similar to *Buzen's algorithm*. The algorithm is considerably simplified by the observation that different customer types do not occur in the relevant summation for the norming constant.

Corollary 5.26 (Norming constant) *Let $S(K, J) = \{(n_1, \ldots, n_J) \in \mathbb{N}^J : \sum_{j=1}^{J} n_j = K\}$. Then*

$$G(K, J) = \sum_{(n_1, \ldots, n_J) \in S(K, J)} \prod_{j=1}^{J} \left\{ \prod_{i=1}^{n_j} \left(\frac{(b_j(\mu_j - 1))}{C(j, i)} \right) \cdot \left(\frac{\mu_j}{\mu_j - 1} \right)^{C(j, n_j)} \right\}.$$

This immediately yields

Theorem 5.27 (Buzen's Algorithm) *(Krüger(83),[89])* *Let $g(k, j), k \geq 0, j \geq 1$ be defined as follows:*

$$g(k, 1) = \prod_{i=1}^{k} \frac{b_1(\mu_1 - 1)}{C(1, i)} \left(\frac{\mu_1}{\mu_1 - 1} \right)^{C(1, n)}, k = 0, 1, \ldots$$

$$g(0, j) = 1, \quad j = 1, 2, \ldots$$

and for $j \geq 2$ and $k > 0$

$$g(k, j) = \sum_{n=0}^{k} \left\{ \prod_{i=1}^{n} \frac{b_j(\mu_j - 1)}{C(j, i)} \left(\frac{\mu_j}{\mu_j - 1} \right)^{C(j, n)} \right\} \cdot g(k - n, j - 1).$$

Then $G(K, J) = g(K, J)$.

Having the steady–state probabilities for the network at hand and an algorithm for computing the norming constant, this enables us to obtain recursively marginal probabilities and mean values of the main performance measures (see [89]). A *Mean Value Algorithm* can be developed similarly to the standard case.

5.7   Open Tandems of Doubly Stochastic Nodes with State Independent Arrival Streams, and Reversibility

In this section we reconsider the modeling principles for networks of doubly stochastic and geometrical nodes as described in definition 5.11, (3) and discussed in the beginning of section 5.3 and in remark 5.23. We determine an important subclass of networks of doubly stochastic nodes where the restrictions put on multiple events of definitions 5.11, (3), are not needed to obtain explicit product form steady states. The starting point is an *output theorem* of Schassberger for the doubly stochastic nodes described in section 5.1 with the additional requirement of state independent Bernoulli arrival process, similar to that of Burke [16] (continuous time exponential nodes) and Hsu and Burke [65] (discrete time state dependent Bernoulli server), further references on classical output results can be found in [75]. From the output theorem we conclude stochastic independence of the queue lengths in equilibrium for open tandems of single server doubly stochastic nodes– without any constraints on simultaneous events at different nodes (which occur as single departures). The concurrent departures occurring at the same time epoch at different nodes are admitted and therefore no restarting of services due to multiple events is necessary. This yields for networks of nodes with service parameter $C_j(n_j) = 1$ for all nodes a product form result without any restart of service or loss of customers at the entrance point of the network.

A similar observation holds for networks of geometrical nodes: For an open tandem of single server Bernoulli nodes with a state independent Bernoulli arrival stream the steady state distribution is of product form, i.e., the joint queue length vector has independent coordinates (see [65] and theorem 4.1) above, which specializes to this case. In equilibrium the departure streams are Bernoulli processes.

Schassberger's result holds even for multiserver doubly stochastic nodes, i.e., for node j we have $C_j(n) > 1$ for some n and we regulate concurrent departures by repititive service/rejection blocking and restart of service ((3) in definition 5.1).

A review on the structure of output processes in continious time and discrete time single node systems can be found in [45]. Using a matrix–analytic approach in this paper for the case of Markov modulated discrete time batch arrival processes and general (discrete) service time distribution in a single server the output process is characterized, allowing further that the server takes occasioally a vacation. Clearly, this prevents us from obtaining independence results similar to those presented here.

Theorem 5.28 (Output theorem) *[117] Consider a doubly stochastic node according to definitions 5.1, 5.2, and 5.5 with state independent Bernoulli-(b) arrival stream of different customer types and state process $X = (X_t : t \geq 0)$ in equilibrium π, see (5.2) of Corollary 5.4, which is in this case*

$$\pi([m(n), k(n); \ldots; m(1), k(1)]) \quad = \quad \left(\frac{b}{c}\right)^n c^{-1} \prod_{h=1}^{n} \frac{a(m(h))}{C(h)} A^{-1}, \tag{5.39}$$

$$[m(n), k(n); \ldots; m(1), k(1)] \in S.$$

Note that we have assumed the D/A−rule and that the state at time t is recorded just after possible departures and arrivals have happened at time t−. Let $D = (D_t : t = 1, 2, \ldots)$ denote the departure process from the node, i.e., $D_t = m$ if at time t− there is a departure of type $m \in M$, and $D_t = 0$ otherwise. Then for any $t_0 \geq 1$ the state X_{t_0} and the departure process $(D_t : t = 1, 2, \ldots, t_0)$ up to time t_0 are stochastically independent. D is an i.i.d. sequence of random variables with

$$P(D_t = m) = ba(m), \quad m \in M, \qquad P(0) = c.$$

Proof : We use induction. By direct computation it is shown that $P(X_1 = x, D_1 = m_1) = P(X_1 = x)P(D_1 = m_1)$ holds. Let $t_0 > 1$.
For $x \in S, m_t \in M \cup \{0\}, t = 1, \ldots, t_0$ we have

$$P(X_{t_0} = x, D_t = m_t, t = 1, \ldots, t_0)$$

$$= \sum_{y \in S} P(X_{t_0} = x, X_{t_0-1} = y, D_t = m_t, t = 1, \ldots, t_0)$$

$$= \sum_{y \in S} P(X_{t_0} = x, D_{t_0} = m_{t_0} | X_{t_0-1} = y, D_t = m_t, t = 1, \ldots, t_0 - 1)$$

$$P(X_{t_0-1} = y, D_t = m_t, t = 1, \ldots, t_0 - 1) \quad =$$

$$\stackrel{(1)}{=} \sum_{y \in S} P(X_{t_0} = x, D_{t_0} = m_{t_0} | X_{t_0-1} = y)$$

$$P(X_{t_0-1} = y | D_t = m_t, t = 1, \ldots, t_0 - 1)P(D_t = m_t, \quad t = 1, \ldots, t_0 - 1)$$

$$\stackrel{(2)}{=} \sum_{y \in S} P(X_{t_0} = x, D_{t_0} = m_{t_0} | X_{t_0-1} = y)\pi(y) \prod_{t=1}^{t_0-1} P(D_t = m_t).$$

Here we used in $\stackrel{(1)}{=}$ the Markov Property of X and in $\stackrel{(2)}{=}$ the induction step. Because $\prod_{t=1}^{t_0-1} P(D_t = m_t) = \prod_{t=1}^{t_0-1} ba(m(t))$, the proof will be done if we can show

$$\sum_{y \in S} P(X_{t_0} = x, D_{t_0} = m_{t_0} | X_{t_0-1} = y)\pi(y) = \pi(x)ba(m_{t_0}). \tag{5.40}$$

We have two cases:

(a) x does not show a fresh customer. Then the predecessors $y \in A_{mi}^0(x), i = 1, \ldots, C(n+1)$, of x which lead by a one−step transition including a departure of a customer of type m to x are described before (5.9) in the proof of theorem 5.3. The LHS of (5.40) is therefore

$$\sum_{i=1}^{C(n+1} \sum_{y \in A_{mi}^0(x)} \left(\pi(x)\frac{ba(m)}{cC(n+1)}\right) cd^-(y, x).$$

(b) x shows exactly one fresh customer located in position $i \in \{1, \ldots, C(n)\}$. Then the predecessors $y \in A^0_{mi}(x), i = 1, \ldots, C(n+1)$, of x which lead by a one–step transition including a departure of a customer of type m to x are described before (5.15) in the proof of theorem 5.3. The LHS of (5.40) in this case is therefore (for i fixed)

$$\sum_{y \in A^0_{mi}} \left(\pi(x) \frac{ba(m)}{ba(m(i))} \right) ba(m(i)) d^+(y, x).$$

$\odot$

Theorem 5.29 *Consider a sequence of doubly stochastic nodes according to definitions 5.1, 5.2, and 5.5. Assume the capacity functions $C_{(\cdot)}$ of the nodes are $C_{(\cdot)}(n) = 1, n \geq 1$. The nodes are numbered 1 through J, customers arrive at node 1 and pass the nodes sequentially in this order, departing from the system after leaving node J. Node 1 is fed by a state independent Bernoulli-(b) arrival stream of different customer types $m \in M$. The state process $X = (X_t = (X_1(t); \ldots, X_J(t)) : t \geq 0)$ with state space S (as described before definition 5.11 but without including the customers' stage on their passage through the tandem) records the customer type sequences at the nodes and the associated residual amount of services requested for. $X_j(t)$ is the local state of node j at time t, $j = 1, \ldots, J$.*

X is positive recurrent if and only if each node in isolation fed by a state independent Bernoulli-(b) arrival stream of different customer types $m \in M$ is positive recurrent. If this is the case then the steady state of X is

$$\pi(x) = \prod_{j=1}^{J} \pi_j(x_j) \quad , \qquad if \quad x = (x_1, \ldots, x_J) \in S, \tag{5.41}$$

where for $j = 1, \ldots, J$ with the notation of (5.27) in corollary 5.14

$$\pi_j(e_j) = c^{-1} A_j^{-1}, \qquad and \tag{5.42}$$

$$\pi_j([m(j, n_j), k(j, n_j); \ldots; m(j, 1), k(j, 1)]) \tag{5.43}$$
$$= \left(\frac{b}{c} \right)^{n_j} c^{-1} \prod_{i=1}^{n_j} a(m(j, i)) A_j^{-1}.$$

Proof : The proof is by induction using theorem 5.28. We assume that X is started with initial distribution $P^{X_0} = \pi$, therefore having independent coordinates at time 0. For $j_0 > 1$ we have

$$P(X_j(t) = x_j : j = 1, \ldots, j_0) \qquad x_j \in S_j, j = 1, \ldots, j_0 \qquad (5.44)$$

$$= \sum_{\overline{m}_{j_0-1} \in (M \cup \{0\})^t} \sum_{\overline{m}_{j_0-2} \in (M \cup \{0\})^t} \cdots \sum_{\overline{m}_1 \in (M \cup \{0\})^t}$$

$$P(X_j(t) = x_j : j = 1, \ldots, j_0, (D_j(1), \ldots, D_j(t)) = \overline{m}_j, j = 1, \ldots, j_0 - 1)$$

$$\overset{(1)}{=} \sum_{\overline{m}_{j_0-1} \in (M \cup \{0\})^t} \sum_{\overline{m}_{j_0-2} \in (M \cup \{0\})^t} \cdots \sum_{\overline{m}_1 \in (M \cup \{0\})^t}$$

$$P(X_{j_0}(t) = x_{j_0} | (D_{j_0-1}(1), \ldots, D_{j_0-1}(t)) = \overline{m}_{j_0-1})$$

$$\cdot \quad P(X_{j_0-1}(t) = x_{j_0-1}, (D_{j_0-1}(1), \ldots$$

$$\ldots, D_{j_0-1}(t)) = \overline{m}_{j_0-1}) | (D_{j_0-2}(1), \ldots, D_{j_0-2}(t)) = \overline{m}_{j_0-2})$$

$$\cdot \quad \ldots$$

$$\cdot \quad P(X_2(t) = x_2, (D_2(1), \ldots, D_2(t)) = \overline{m}_2) | (D_1(1), \ldots, D_1(t)) = \overline{m}_1)$$

$$\cdot \quad P(X_1(t) = x_1, (D_1(1), \ldots, D_1(t)) = \overline{m}_1)$$

In $\overset{(1)}{=}$ we used successively that due to the single server property $C_j(n) = 1, n \geq 0$, and due to the linear structure of the system $(X_j(t) = x_j, (D_j(1), \ldots, D_j(t)) = \overline{m}_j)$ is independent of $(X_i(t) = x_i : i = 1, \ldots, j - 2, (D_i(1), \ldots, D_i(t)) = \overline{m}_i, i = 1, \ldots, j - 2)$ given $(D_{j-1}(1), \ldots, D_{j-1}(t)) = \overline{m}_{j-1})$ and $X_j(0)$.

Applying theorem 5.28 to (5.44) yields

$$P(X_j(t) = x_j : j = 1, \ldots, j_0)$$

$$= \sum_{\overline{m}_{j_0-1} \in (M \cup \{0\})^t} \sum_{\overline{m}_{j_0-2} \in (M \cup \{0\})^t} \cdots \sum_{\overline{m}_1 \in (M \cup \{0\})^t}$$

$$P(X_j(t) = x_j : j = 1, \ldots, j_0, (D_j(0), \ldots, D_j(t)) = \overline{m}_j, j = 1, \ldots, j_0 - 1)$$

$$= \sum_{\overline{m}_{j_0-1} \in (M \cup \{0\})^t} \sum_{\overline{m}_{j_0-2} \in (M \cup \{0\})^t} \cdots \sum_{\overline{m}_2 \in (M \cup \{0\})^t}$$

$$P(X_{j_0}(t) = x_{j_0} | (D_{j_0-1}(0), \ldots, D_{j_0-1}(t)) = \overline{m}_{j_0-1})$$

$$\cdot \quad P(X_{j_0-1}(t) = x_{j_0-1}, (D_{j_0-1}(0), \ldots$$

$$\ldots, D_{j_0-1}(t)) = \overline{m}_{j_0-1}) | (D_{j_0-2}(0), \ldots, D_{j_0-2}(t)) = \overline{m}_{j_0-2})$$

$$\cdot \quad \ldots$$

$$\cdot \quad P(X_2(t) = x_2, (D_2(0), \ldots, D_2(t)) = \overline{m}_2)) P(X_1(t) = x_1)$$

Applying now the induction hypthesis we obtain the independence of $(X_j(t) : j = 1, \ldots, j_0)$. $\odot$

Note that the result and the proof of theorem 5.12 do not apply to the situation and framework of theorem 5.29 because we did not impose the restriction (3) of definition 5.11 on multiple events originating in joint events from different nodes. Multiple events due to a departure at some different nodes, combined with an arrival there are allowed, because there is at most one departure at every time instant at a single node. So only *(D)* of definition 5.2 applies and there are no multiple departure candidates at each node.

Corollary 5.30 *Consider a sequence of single server queues in discrete time with a state independent Bernoulli-(b) arrival stream of customers with different customer types. The service requests of the customers may be node and type dependent and constitute an independent family of random variables which is independent of the arrival decisions. An arriving customer is of type $m \in M$ with probability $a(m)$ and requests at node j for k_j quanta of service time with probability $p_{j,m}(k_j), k_j \geq 1$.*
If the service regime of the nodes is either Last–Come–First–Served–Preemptive–Resume *or* Random Service Allocation *or* Round–Robin–Premptive–Resume *(for details see [32]) then in equilibrium the nodes behave as if they are independent. The steady state distribution is given by*

$$\pi([m(j, n_j), k(j, n_j); \ldots; m(j, 1), k(j, 1)] : j = 1, \ldots, J) \qquad (5.45)$$

$$= \prod_{j=1}^{J} \left\{ \left(\frac{b}{c} \right)^{n_j} c^{-1} \prod_{i=1}^{n_j} \left(a(m(j, i)) \sum_{h_j = k_j}^{\infty} p_{j, m(j,i)}(h) \right) A_j^{-1} \right\}.$$

The steady state joint queue length distribution is insensitive with respect to the shape of the service time distribution as long as their means are not perturbed.

Corollary 5.31 *Consider a sequence of doubly stochastic queues according to definitions 5.1, 5.2, and 5.5 in discrete time with a state independent Bernoulli-(b) arrival stream of customers with different customer types. For each node internal concurrent departure requests of departure candidates at this node are regulated according to repetitive service/rejection blocking with restarting the jobs of the departure candidates according to definition 5.1, part (3). Exception: when there is an additional arrival at this node one departure candidate is selected at random to depart immediately, substituted in his position by the arriving customer. Single departures realised at the same time epoch at different nodes are admitted.*
The service requests of the customers may be node and type dependent and constitute an independent family of random variables which is independent of the arrival decisions. The state process X is positive recurrent if and only if each node in isolation fed by a state independent Bernoulli-(b) arrival stream of different customer types $m \in M$ is positive recurrent. If this is the case then the steady state of X is of product form and given by (5.41).

The proof of corollary 5.31 is similar to that of theorem 5.29. For the case of having only one customer class, (i.e., indistinguishable customers with identical stochastic behaviour traverse the tandem,) there is another way to prove the theorems and corollaries of this section. For this we notice that the results in theorem 5.28 imply that the doubly stochastic queue in isolation is quasi reversible in the sense of definition 12.6 in [18], which can be concluded from lemma

12.8 there. As a consequence for the case of indistinguishable customers traversing the linear tandem, theorem 12.15 in [18] yields the results of theorem 5.40 and corollary 5.42 for that case.

It should be noticed that some care is needed in applying the quasi reversibility theorems from [18]. This originates from the different regulation schemes for the arrival and departure events which are prescribed, see figures 2.2 and 2.3. The definition of quasi reversibility requires that the state of the node is observed at epochs between the possible departure and arrival in $[t, t+1)$, p. 350 in [18]. Nevertheless the qualitative characterization of quasi reversibility via lemma 12.8 in [18] meets the properties obtained in the output theorem 5.28. Another way to resolve the problem is to reconsider the pathwise construction of the state process for the single nodes in corollary 2.2.

It should be further remarked that connecting these nodes to a tandem and then comparing the models and applying the proved theorems vice versa possibly may result in slight problems: In the framework of [18] the customers depart from a node at the beginning of a time slot (i.e., at time $t+$) and arrive at the next station at the end of the same time slot $[t, t+1)$ (i.e., at $(t+1))-$). So transition needs a unit of time, while in our models transition is assumed to be immediate. It follows that although the state processes observed at slot boundaries are stochastically indistinguishable, care is needed in drawing conclusions about derived performance measures, e.g. on mean passage times via Little's formula.

Chapter 6

General Networks with Batch Movements and Batch Services

The structural and computational simplicity of classical product form queueing networks in continuous time (see e.g.: [75], [4], [69], [50]) relies on the property that independently moving customers which are independently served with probability one do not jump at the same time instant. From about 1990 on a sequence of papers appeared which considered explicitly networks where servicing is done for customer batches and where customers are forwarded into and through the network in batches. Before that, similar problems were mostly dealt with via simulation or using approximations. Only special mini networks were solved explicitly.

Now there is a formalism and computational approach to solve classes of queueing models where explicit steady–state probabilities can be given although batch movements and batch services are allowed, a review on the state of the art can be found in [18] and [121].

In discrete time networks, phenomena like batch processing and travelling in batches are natural features of the systems, as discussed in chapter 4. Fortunately enough with respect to discrete time networks, most of the underlying principles developed for continuous time batch movements networks carry over directly. In most of the papers to be mentioned below the results were therefore derived for both cases simultaneously. The different interpretations of the formulas – which indeed are necessary for correctly applying them – have to be done carefully. (See Remark 6.8.)

Reviewing the literature we see that a great number of different, but related models were developed. Indeed, some inventions were done in parallel, but often the explicit notation differs considerably. In the book of Chao, Miyazawa, and Pinedo [18] a considerable part of general network models are described. The book mainly deals with continuous time systems, but the formalism is versatile

H. Daduna: Queueing Networks with Discrete Time Scale, LNCS 2046, pp. 105–120, 2001.
© Springer-Verlag Berlin Heidelberg 2001

enough to carry over to discrete time scale, chapter 12 of the book is devoted to that.

Naturally, this chapter has some overlap with the contents of chapter 12 of [18], because some of the underlying references coincide, but on the other hand there are sufficient many differences, originating in different points of view on the field. The aim of this chapter is, to compare the systems of chapters 2 to 5 with the general batch movement sytems from the literature.

Sections 6.1 and 6.4 describe a rather general model and its steady state behaviour, in section 6.5 we discuss special examples, in section 6.5 the networks introduced by Walrand [135], and finally we sketch additional triggering for special events in the general network.

6.1   The General Network Model

The central point of this section will be a short desription and discussion of a model which in my opinion is an appealing representative of what is going on in the field. The presentation follows Henderson and Taylor [56], [58], Miyazawa [98], [99], and Ozawa [104]. A recent discussion of structural properties of these models can be found in [23]. (This paper contains references for further applications of the models, e.g., to Petri nets.) Partially parallel developments including some additional features in the networks as, e.g., state dependent routing are presented by Boucherie and van Dijk [8], [6].

The general formalism is rather abstract and therefore widely applicable. I shall then describe a sequence of prototype models, from the literature in some detail. This shows that it is often not an easy task to fit a concrete example into the formalism. Some related models procedures for special applications may be found in Woodward's book [142].

We consider an open network of queues with nodes numbered $1, 2, \ldots, J$. Customers enter the system from the outside (which we usually consider as node 0), procede according to some routing regime through the network and eventually leave the system. The customers may be of different types which they may randomly change when entering a new node. The set M of customer types is assumed to be finite.

The system evolves in discrete time $I\!N$ according to some Markovian transition law. We denote the state process of the system by $X = (X_t : t \in I\!N)$, living on the state space S which is $(I\!N^{|M|})^J$ or a subset thereof.

X carries the following detailed information: $X = (X_t : t \in I\!N) = ((X_t(j, m) : j = 1, \ldots, J, m \in M) : t \in I\!N)$, where $X_t(j, m) = n(j, m)$ indicates that at time t there are $n(j, m)$ customers of type m present at node j.

The development of X is governed by sequences of *release vectors* $D = (D_t : t \in I\!N)$ and *transformed vectors* $A = (A_t : t \in I\!N)$. For $t \in I\!N$ we have release vector

$$D_t = (D_t(0), D_t(j, m), 1 \le j \le J, m \in M),$$

and transformed vector

$$A_t = (A_t(0), A_t(j, m), 1 \le j \le J, m \in M).$$

Their functioning and interplay is as follows:

Assume at time $t \in \mathbb{N}$ the network to be in state $X_t = n$. Then at time $(t + 1)-$ from the source (node 0) $D_t(0) = a(0)$ customers are released for being transformed into the network. From node j there are $D_t(j, m) = a(j, m)$ customers of type m released to be transformed to some other node or to the sink (node 0). Immediately thereafter (*between time* $(t+1)-$ *and time* $t+1$) the released customers are transformed – possibly changing their types – to their destination nodes: $A_t(0) = a'(0)$ customers depart from the network, $A_t(j, m) = a'(j, m)$ customers of type m enter node j, $1 \le j \le J, m \in M$. Updating the state of the network according to this movements and changes we obtain X_{t+1}. Formally: For

$$a = (a(0), a(j, m), 1 \le j \le J, m \in M)$$

let

$$a^+ = (a(j, m), 1 \le j \le J, m \in M),$$

and similarly let A_t^+ and D_t^+ be obtained from A_t and D_t by deleting the *external* components $A_t(0)$ and $D_t(0)$. Then

$$X_{t+1} = X_t - D_t^+ + A_t^+, \quad t \in \mathbb{N}, \tag{6.1}$$

and

$$X_t \ge D_t^+, \quad and \quad X_{t+1} \ge A_t^+, \tag{6.2}$$

where the inequalities here and in the following are to be read coordinatewise. Having now

$$X_{t+1} = n', X_t = n, D_t = a, A_t = a' \quad then \quad n' = n - a^+ + a'^+.$$

Because new customers arrive only from the outside and vanishing customers must depart to the sink (node 0) we have the balance equation

$$D_t(0) + \sum_{j=1}^{J} \sum_{m \in M} D_t(j, m) = A_t(0) + \sum_{j=1}^{J} \sum_{m \in M} A_t(j, m). \tag{6.3}$$

The sequences A and D therefore assume only values from a commom state space which we denote by $\mathcal{A} \subseteq \mathbb{N} \times \mathbb{N}^{J \cdot |M|}$. It should be noticed that the images of A and D nevertheless need not be identical.

The following probabilistic assumption on A and D imply that X as given by (6.1) is a Markov chain with stationary transition probabilities:

For $t \in \mathbb{N}$ D_t depends on the history of the system up to time t only through X_t and

$$P(D_t = a | X_t = n, X_s = n_s, D_s = a_s, A_s = a'_s, 0 \le s < t) \tag{6.4}$$
$$= P(D_t = a | X_t = n) = q(n, a), \quad n \in S, a \in \mathcal{A}, \quad with \quad (6.2),$$

A_t depends on the history of the system up to time $(t+1)-$ only through D_t and

$$P(A_t = a'|D_t = aX_s = n_s, D_s = a_s, A_s = a'_s, 0 \leq s < t) \qquad (6.5)$$
$$= \quad P(A_t = a'|D_t = a) = r(a, a'), \quad a, a' \in \mathcal{A}, \quad \text{with} \quad (6.3).$$

The sequence of transformations of released vectors into transformed vectors governed by (6.5) shall be called *routing process*. The transition probabilities of X are

$$P(X_{t+1} = n'|X_t = n) = \sum_{\substack{a, a' \in \mathcal{A} \\ n - a^+ + a'^+ = n'}} q(n, a)r(a, a'), \quad n, n' \in S.$$

From (6.3) it follows that a Markov chain on $\mathcal{A}$ determined by the transition matrix $R = (r(a, a') : a, a' \in \mathcal{A})$ (the routing matrix) is not irreducible. Vectors a and a' connected by (6.3) are assumed to be reachable vice versa via R. We assume

Assumption 6.1 *For the transition matrix* $R = (r(a, a') : a, a' \in \mathcal{A})$ *on* $\mathcal{A}$ *denote by* $\mathcal{A}_k \subseteq \mathcal{A}$ *the set of those states in* $\mathcal{A}$ *which contain exactly k customers,* $k \in \mathbb{N}$. *The sets* $\mathcal{A}_k$ *are finite and we assume henceforth that R restricted to each* $\mathcal{A}_k$ *is irreducible.*

6.2   Steady States

The essential prerequisits for the successful steady–state analysis are the following assumptions. These assumptions or similar ones can be found in almost all papers on the subject which we consider here.

Assumption 6.2 *There exists a strict positive function* $\Phi : S \longrightarrow (0, \infty)$, *and non negative functions* $\Psi : S \times \mathcal{A} \longrightarrow [0, \infty)$, *and* $\Theta : \mathcal{A} \longrightarrow [0, \infty)$, *such that for all* $n \in S$ *and all* $k \in \mathbb{N}$ *the partial functions*

$$\Psi(n, \cdot) : \mathcal{A}_k \longrightarrow [0, \infty) \qquad (6.6)$$
$$a \to \Psi(n, a)$$

are constants, and

$$q(n, a) = \frac{\Psi(n - a^+, a) \cdot \Theta(a)}{\Phi(n)}, \quad n \in S, \quad a \in \mathcal{A}. \qquad (6.7)$$

Assumption 6.3 *There exists strict positive functions* $f : \mathcal{A} \longrightarrow (0, \infty)$, *and* $g : S \longrightarrow (0, \infty)$, *such that* f *solves the following* traffic equations *for batch movement systems*

$$\Theta(a)f(a) = \sum_{a' \in \mathcal{A}} \Theta(a')f(a')r(a', a), \quad a \in \mathcal{A}, \qquad (6.8)$$

and there is a representation

$$\frac{g(n)}{g(n - a^+ + a'^+)} = \frac{f(a)}{f(a')} \quad \forall n \in S, a, a' \in A \quad \text{with} \quad q(n,a)r(a,a') > 0. \quad (6.9)$$

Theorem 6.4 (Steady state) *Suppose that assumptions 6.1, 6.2, 6.3 hold. Then the state process X of the network has an invariant measure $\hat{\pi}$ on S given by*

$$\hat{\pi}(n) = \Phi(n)g(n), \quad n \in S. \quad (6.10)$$

$\hat{\pi}$ can be normalized to an invariant probability law π of X on S if and only if $C = \sum_{n \in S} \Phi(n)g(n) < \infty$. If this is the case we have

$$\pi(n) = C^{-1}\Phi(n)g(n), \quad n \in S. \quad (6.11)$$

Proof: Application of time reversing arguments according to proposition 7.2 of section 7.1 of the appendix (see [56] and [99]):

We expect that the time reversal $\bar{X}$ of the state process X describes the evolution of a system with a similar structur as the original system. The release probability $\bar{q}(n, a)$ similar to (6.4) is the same as in the original system and given by (6.7) as well, while the routing probability $\bar{r}(\cdot, \cdot)$ similar to (6.5) is given by

$$\bar{r}(a', a) = \frac{\Theta(a)f(a)r(a,a')}{\Theta(a')f(a')} \quad a', a \in A,$$

According to proposition 7.2 of section 7.1 we have to prove for the non normalized measure $\hat{\pi}$ and states $n, n' \in S$

$$\hat{\pi}(n) \sum_{\substack{a,a' \in A \\ n-a^+ + a'^+ = n'}} q(n,a)r(a,a') = \hat{\pi}(n') \sum_{\substack{a',a \in A \\ n'-a'^+ + a^+ = n}} q(n',a')\bar{r}(a',a). \quad (6.12)$$

Inserting (6.10) and (6.7) into (6.12) and then applying (6.9) yields for the lefthand side of (6.12):

$$\Phi(n)g(n) \sum_{\substack{a,a' \in A \\ n-a^+ + a'^+ = n'}} \frac{\Psi(n-a^+,a)\Theta(a)}{\Phi(n)}r(a,a') \quad (6.13)$$

$$= \sum_{\substack{a,a' \in A \\ n-a^+ + a'^+ = n'}} \Psi(n-a^+,a)\Theta(a)g(n-a^+ + a'^+)\frac{f(a)}{f(a')}r(a,a').$$

The righthand side of (6.12) is

$$\Phi(n')g(n') \sum_{\substack{a',a \in A \\ n'-a'^+ + a^+ = n}} \frac{\Psi(n'-a'^+,a')\Theta(a')}{\Phi(n')}\bar{r}(a,a')$$

$$= \sum_{\substack{a',a \in A \\ n'-a'^+ + a^+ = n}} \Psi(n'-a'^+,a')\Theta(a')g(n')\frac{\Theta(a)f(a)r(a,a')}{\Theta(a')f(a')}. \quad (6.14)$$

Now for each pair $a, a' \in \mathcal{A}$ such that $r(a, a') > 0$ with $n - a^+ + a'^+ = n'$ the summands in (6.13) and (6.14) balance separately because $\Psi(n - a^+, a) = \Psi(n' - a'^+, a')$. The latter holds because $n - a^+ = n' - a'^+$ and a, a' are in the same class $\mathcal{A}_k$ (see assumption 6.2).   ⊙

For an explicit evaluation of π according to (6.11) we need an explicit expression for $g(n), n \in S$, which is easier accessible than the *representation* given in (6.9) which in general is not easy to handle.

For the case that in state $n \in S$ a transition to $n' = n - a^+ + a'^+$ is possible such that $a^+ - a'^+ > 0$ holds, (6.9) suggests a recursion step of the form

$$g(n) = g(n - a^+ + a'^+) \cdot \frac{f(a)}{f(a')} \quad , \tag{6.15}$$

and provided this iteration can be continued we may hope to end eventually by $g(\tilde{n})$, for some base state $\tilde{n}$, which in open networks usually can be chosen as $\tilde{n} = 0$, the empty network. Provided further, that this iteration for all states can be done in a *well defined* way, we would be able to compute the network's equilibrium.

The term *well defined* means that the result obtained by the iterative procedure does not depend on how the path from the base state $\tilde{n}$ to any other state n is chosen in applying (6.15).

As Miyazawa [98] noticed, almost all examples of open batch movement networks which can be found in the literature, where an explicit product form equilibrium is known, show the following structure: The base state is the empty state of the network, $\tilde{n} = 0$. Let $e_0 \in \mathcal{A}$ denote the unit vector having 1 in the 0th coordinate and other coordinates 0, and $e_{i,m} \in \mathcal{A}$ the unit vector having 1 in the (i, m)th coordinate and other coordinates 0. If $q(n, e_0) r(e_0, e_{i,m}) > 0$ then (6.9) and (6.15) imply

$$g(n + e_{i,m}^+) = g(n) \cdot \frac{f(e_{i,m})}{f(e_0)}.$$

Denoting

$$\frac{f(e_{i,m})}{f(e_0)} =: \alpha_{i,m}, \quad i = 1, \ldots, J, m \in M,$$

we obtain

$$g(n) = g(0) \prod_{j=1}^{J} \prod_{m \in M} \alpha_{j,m}^{n(j,m)}, \quad n \in S,$$

$$f(a) = f(e_0) \prod_{j=1}^{J} \prod_{m \in M} \alpha_{j,m}^{a(j,m)}, \quad a \in \mathcal{A}.$$

For a discussion of the structural properties of functions obeying such a constructive representation see [120], p.149 and 155. Although the derivation there is done for continuous time systems the conclusions are relevant for our context as well. Note that Serfozo [120] discusses closed network applications as well.

Remark 6.5 *In [56] and [98] (and the references given there) a stronger condition than assumption 6.2 is required. It is assumed that* $\Psi : S \times \mathcal{A} \longrightarrow [0, \infty)$, *is a function of S only, independent of the second coordinate $\mathcal{A}$. In [99] it was remarked that (6.6) suffices to prove the theorem. The interpretation is:*

For $a, b \in \mathcal{A}$ we write $a \rightleftharpoons b$ if and only if a, b are members of the same communicating class with respect to the transition matrix R of the routing process. Then $\Psi(\cdot, \cdot)$ depends in its second coordinate only through the equivalence classes with respect to $\rightleftharpoons$. So Ψ can be written as $(n, a) \longrightarrow \Psi(n, a) := \tilde{\Psi}(n, a/\rightleftharpoons)$, where

$$\tilde{\Psi} : S \times (\mathcal{A}/\rightleftharpoons) \longrightarrow [0, \infty)$$

is a function defined in its second coordinate on the equivalence classes of $\mathcal{A}$ modulo $\rightleftharpoons$.

The introduction of the additional dependency of Ψ on a second coordinate with respect to $\mathcal{A}$ is assumed to broaden the applicability of the concept.

Remark 6.6 *In the definition (6.7) of the service allocation function $q(\cdot, \cdot)$ the functions Ψ, Φ, θ cannot be chosen arbitrarily, because for any n the function $q(n, \cdot)$ is a counting density on $\mathcal{A}$. $\Phi(n)$ is therefore the norming constant of $q(n, \cdot)$. In continuous time this restriction does not hold because $q(\cdot, \cdot)$ is a rate then and normalization is not necessary.*

The form of $q(\cdot, \cdot)$ allows a versatile modeling of service features. E.g., the total number of customers moving in one time step can be bounded by setting $\Theta(a) = 0$ if $a(0) + \sum_{j=1}^{J} \sum_{m \in M} a(j, m) > B$ for a prescribed bound B. In example 6.11 this technique is used to obtain networks with single arrivals.

6.3 Related Examples and Discussion of the Model

In the literature a lot of companion models to that considered in section 6.1 can be found. Most of them describe systems which evolve in continuous time. Some representatives will be sketched in the following. Further models and references can be found in the references notes in [18] (section 12.9, p. 403).

Remark 6.7 (Local balance and reversibility) *Boucherie and van Dijk [7], [8] used a slightly different formalism to investigate queueing network models with batch movements, both in continuous and in discrete time. They introduced special local balance features observed in these networks to break the global balance equations into subequations which results in product form structures of the solution.*

In [8] the result of Theorem 6.3 is generalized by incorporating state dependencies into the routing schemes of the customers. These dependencies are subject to an appropriate balancing behaviour.

Serfozo [120] developed a product form equilibrium formalism for queueing networks with concurrent (batch) movements in continuous time. The key property of the network processes (joint queue length vectors) is time reversibility. Applying the structural properties provided by Kolmogorov's criteria this yields product form equilibrium. It turns out that there is a close connection to multidimensional birth–death–processes and to migration processes with concurrent migrations.

The relevance of the continuous time network models to our discrete time framework is at least twofold. We obtain two different approaches to discrete time modeling. These provide us with a simple machinery to produce discrete time stochastic networks from the more familiar continuous time counterparts by simply reinterpreting the models' description and data.

First we may consider the embedded jump chain of the continuous time network process. This is a discrete time Markov chain describing a batch movements process where time durations are not taken into consideration. Substituting the exponential clocks by a synchronized clock with unit time steps we obtain a generic discrete time queueing network with batch movements of the customers.

Second, assuming that the transition rates of the networks are bounded, we may apply a uniformization procedure to the network process. (See [73], chapter 2.) This yields a continuous time Markov process which is stochastically indistinguishable from the original process. The times between jumps are governed by a homogeneous Poison process with dummy jumps inserted, where the network jumps without time lag out of and into the same state. Considering the embedded jump chain (including the dummy jumps) yields a discrete time Markov chain with batch customer movements showing the same steady state as the original continuous time process. Here the interpretation is that the homogeneous Poisson clock is substituted by a synchronized clock with unit time steps.

Remark 6.8 (Scheduling of customers in the nodes) *The state process under consideration only counts for the number of customers of different classes present at the nodes. Queueing due to an internal queueing discipline of the nodes can not be dealt with in general. Therefore the following physical interpretation of the internal node behaviour might be helpful for applying the abstract formalism in a modeling process:*
The nodes operate as infinite server queues with deterministic service time requests of one time unit for the customers. The service allocation is state–dependent according to $q(\cdot,\cdot)$, see (6.4).

Reconsidering the generalized form (6.7) of $q(\cdot,\cdot)$, this interpretation confirms the statement of Remark 6.5: Without the introduction of the second coordinate of Ψ in (6.6) the dependence of the service allocation function $q(n,a)$ on the released batch a and the stay–on customer vector $n - a^+$ would be of product form. This restrictive assumption is weakend by assuming (6.6).

Taking into account explicit queueing behaviour by considering customer positions is possible but needs restricted assumptions on how newly arriving customers are inserted into the queues. For the case of (similarly defined) closed networks of queues in [60] it is assumed that the queues are organized according to the shift–protocol (see the description in Section 5.1) and that newly arriving customers choose their position at random from the feasible positions. This is the same arrival regime as described in (2) of definition 5.1 of Section 5.1 for doubly stochastic nodes.

If all customers in the network are of the same type we can assume that any service discipline is applied for selecting those customers who obtain the service capacity which is provided to the node according to $q(\cdot,\cdot)$. Especially we may assume that service is provided according to FCFS (first–come–first–served) as we have done in chapter 3 and 4.

Remark 6.9 (Closed networks) We have described the network up to now as an open system where customers enter from the outside, travel through the system and eventually leave the system. It is easy to carry over the development to the case of closed networks. We fix a constant number K of customers cycling in the network of nodes $1,\ldots,J$ and changing their types according to $q(\cdot,\cdot)$ and $r(\cdot,\cdot)$, where the zero–coordinate is omitted. The release vectors then are of the form $D_t = (D_t(j,m), 1 \le j \le J, m \in M)$, and transformed vectors are of the form $A_t = (A_t(j,m), 1 \le j \le J, m \in M)$. For details see [60].

Some examples will now be discussed which are of special interest for applications. From a historical point of view they can be considered as landmarks on the way of developing discrete time stochastic network theory. We show examples for network nodes with internal queueing structure as well as for nodes without such specification.

Example 6.10 (Networks with independent routing of customers)
(See [56] and for the closed network case [60].) This is a network with the standard routing process used in the most applications up to now. Independent Markovian routing governs the customers' movements in the classical continuous time networks of Jackson [69] and Gordon and Newell [50], and in the BCMP–networks ([4].

The routing process is as follows: A customer of type m departing from node i jumps to node j and changes to type m' with probability $p(i, m; j, m')$, independent of anything else in the network. This routing neither depends on the composition of the rest of the release vector nor on the composition of the rest of the transformed vector. For $j = 0$ this is the probability that a customer of type m on leaving node j departs from the network as a customer of type m'. By $p(0; i, m)$ we denote the probability that a customer from the outside enters the network as a type m customer joining node i. This shall happen independent from the history of the network.

The routing process defined by p on $\mathcal{A}_k$, can be interpreted as a migration process for k individuals on a set of $J \cdot |M| + 1$ nodes, $k \in \mathbb{N}$. Considering only one single customer on that network his steady state behaviour

$y = (y(0), y(j, m) : 1 \le j \le J, m \in M)$ *will be determined by the* traffic equation

$$y(i, m) \quad = \quad y(0)p(0; i, m) + \sum_{j=1}^{J} \sum_{m' \in M} y(j, m')p(j, m'; i, m), \qquad (6.16)$$

$$1 \le i \le J, \quad m \in M.$$

Under our overall assumption 6.1 of irreducibility for the routing process on the components A_k an invariant measure for the routing process (taking into account all the possible permutations of the independently travelling customers) is

$$x(a) = \frac{y(0)^{a(0)}}{a(0)!} \prod_{j=1}^{J} \prod_{m \in M} \frac{y(j, t)^{a(j, m)}}{a(j, m)!}, \quad a \in \mathcal{A}$$

We assume in this example the release probability $q(n, a)$ to be inversely proportional to $a(0)!$ and $a(i, m)!, 1 \le i \le J, m \in M$. This is granted e.g. by taking with $C > 0$

$$\Theta(a) = \frac{C}{a(0)! \prod_{j=1}^{J} \prod_{m \in M} a(j, m)!}, \quad a \in \mathcal{A}. \qquad (6.17)$$

It turns out that for any network having the form (6.17) for Θ and having independent routing, suitable functions f, g are

$$f(a) \quad = \quad \prod_{j=1}^{J} \prod_{m \in M} y(j, t)^{a(j, m)}, \quad a \in \mathcal{A},$$

$$g(n) \quad = \quad \prod_{j=1}^{J} \prod_{m \in M} y(j, t)^{n(j, m)}, \quad n \in S.$$

Thus from theorem 6.4 we conclude that if the network can be stabilized then its steady state probability is given as

$$\pi(n) = C\Phi(n) \prod_{j=1}^{J} \prod_{m \in M} y(j, t)^{n(j, m)}, \quad n \in S,$$

where $\Phi(\cdot)$ is the norming constant for the release probabilities.

Example 6.11 (Open tandem systems with geometrically distributed batches moving) *Harnpanichpun and Pujolle [54] applied the result of Henderson and Taylor (see Theorem 6.4) to sequentially ordered nodes $1, \ldots, J$, where indistinguishable customers enter the network at node 1, proceed to node $2, \ldots,$ and eventually depart from the network via node J. The regulation of the movements is assumed to be of the form as desribed above: departures just before arrivals, and arrivals being late at times $t-, t \in \mathbb{N}$. The network's state, recording the joint queue length vector X_t is observed at times $t, t \in \mathbb{N}$.*

The external arrivals occur at node 1 only: at each time instant there will arrive a batch of k customers with probability $(1-\lambda)\lambda^k$, $k \in \mathbb{N}$, for some $\lambda \in (0,1)$. The sequence of batch sizes is assumed to be i.i.d. and to be independent from anything else. The departure probabilities are node specific and depend on the history of the system only through the actual state of the node under consideration:

If at node j there are n_j customers present at time t then with probability $\mu_j^k/(1 + \mu_j + \mu_j^2 + \cdots + \mu_j^{n_j})$ at time $(t+1)-$ a batch of size k will leave that node, $(\mu_j \in (0,1))$, for all nodes $j = 1, \ldots, J$.

The Markovian joint queue length process $X = (X_t : t \in \mathbb{N})$ is ergodic if and only if $\lambda < \mu_j$, $j = 1, \ldots, J$. The steady state distribution is

$$\pi(n_1, \ldots, n_J)$$
$$= (1-\lambda)^J \prod_{j=1}^{J} \left\{ \left(1 - \frac{\lambda}{\mu_j}\right) \left(\frac{\lambda}{\mu_j}\right)^{n_j} \left(1 + \mu_j + \mu_j^2 + \cdots + \mu_j^{n_j}\right) \right\},$$
$$n_j \in \mathbb{N}, j = 1, \ldots, J.$$

(For the details of how to apply the formalism of Theorem 6.4 see [54].)

The linear structure of the network seems to be essential for the result described here. For a discussion and some counterexamples on nonlinear networks with truncated geometrically distributed batch service see [59].

In both examples 6.10 and 6.11 the internal service regime of the nodes is on a batch size basis: In example 6.10 according to (6.17) we have (nearly) truncated Poisson distributed batches released, while in example 6.11 the released batches are truncated geometrically distributed.

The next example contributes to the discussion of batch movement by reconsidering individual queues with single customer arrivals and services. The batch movements nevertheless occur from the synchronized network behaviour due to the discrete time scale. This is a subclass of the models in chapter 4. We demonstrate here how this subclass fits into the formalism of section 6.1.

Example 6.12 (Open tandems of state dependent Bernoulli servers)

Hsu and Burke [65] considered an open tandem of nodes $1, \ldots, J$ where a Bernoulli process with arrival probability $b \in (0,1)$ feeds node 1 with a stream of undistinguishable customers who proceed stepwise through the sequence of nodes. The service probabilities are node specific and depend on the history of the system only through the actual queue length at the respective nodes.

If at time $t \in \mathbb{N}$ there are $n_j > 0$ customers present at node j than there is with probability $p_j(n_j)$ exactly one depature from j at time $(t+1)-$, with probability $1 - p_j(n_j)$ all customers stay at node j for at least one more time unit, $p_j(0) = 0$. If the system is ergodic its unique steady state is (see theorem 4.1)

$$\pi(n_1, \ldots, n_J) = C \prod_{j=1}^{J} \left\{ \prod_{h=1}^{n_j} \frac{b(1 - p_j(h-1))}{(1-b)p_j(h)} \right\}, \quad n_j \in \mathbb{N}, j = 1, \ldots, J.$$

As Henderson and Taylor (see Section 3.4 in [56]) noticed, this can be proved by an application of Theorem 6.4. They define for $n = (n_1, \ldots, n_J) \in \mathbb{N}^J$ and $a = (a_0, a_1, \ldots, a_J) \in \mathbb{N}^{J+1}$

$$\Theta(a) = \begin{cases} b^{a_0}(1-b)^{(1-a_0)} & if \quad \max_{0 \leq j \leq J}(a_j) \in \{0, 1\} \\ 0 & otherwise \end{cases} ,$$

$$\Psi(n - a^+) = \prod_{j=1}^{J} \left\{ (1 - p_j(n_j - a_j)) \prod_{h=1}^{n_j - a_j} \left(\frac{1 - p_j(h-1)}{p_j(h)} \right) \right\},$$

$$\Phi(n) = \prod_{j=1}^{J} \prod_{h=1}^{n_j} \left(\frac{1 - p_j(h-1)}{p_j(h)} \right),$$

and obtain for n, a as above

$$q(n, a) = b^{a_0}(1-b)^{(1-a_0)} \prod_{j=1}^{J} \left\{ (1 - p_j(n_j))^{(1-a_j)} p_j(n_j)^{a_j} \right\}.$$

The functions f, g defined for n, a as above are

$$f(a) = \left(\frac{1-b}{b} \right)^{a_0}, \qquad g(n) = \left(\frac{b}{1-b} \right)^{n_1 + \cdots + n_J}.$$

A property which is common to the systems of examples 6.11 and 6.12 is that the external arrivals are governed by *batch Bernoulli processes* ([18], definition 12.5). These are simply $\mathbb{N}$ valued sequences of identically distributed independent random variables $Z = (Z_t : t \in \mathbb{N})$, where Z_t denotes the number of arrivals in $[t, t+1)$. Z_t is assumed to be independent of the history of the system up to and including t. In example 6.11 the Z_t are geometrically distributed on $\mathbb{N}$ with parameter $1 - \lambda$, in example 6.12 Z_t is Bernoulli(b) distributed on $\{0, 1\}$. A further example will be provided by Walrand's S–queue to be presented in the next section, where Z_t is Poisson(λ) distributed. In all cases it can be proved that in connection with the selected service regime the nodes transform the batch Bernoulli arrival stream into a batch Bernoulli departure stream with the same stochastic behaviour. Moreover all of these nodes are quasi reversible according to definition 12.6 in ([18]. It follows that these systems can be coupled to a linear tandem system with product form steady state. Further examples for similar procedures are given in corollaries 5.30 and 5.31, compare the discussion below corollary 5.31.

Example 6.13 *A class of systems different from those dealt with in examples 6.13 – 6.12 are stochastic Petri nets with product form equilibrium. A direct application of theorem 6.4 to solve a Petri net model can be found in [57]. For further relevant literature see the references in [55].*

6.4 Walrand's S–Queues and Networks

Walrand introduced a network of so–called S–queues in discrete time and proved that with an arrival sequence of independent Poisson–distributed customer batches and with a specific release probability the isolated nodes are quasi–reversible [135]. From this observation it follows that these nodes can be used as building blocks of networks with a product form equilibrium. For more detailed investigation of S–queues see [142] and [18]. For a single node Walrand's result is:

Theorem 6.14 *Let $A = (A_t : t \in I\!N)$ be an $I\!N$–valued i.i.d. arrival sequence of Poisson–λ variables (the transformed variables) and $D = (D_t : t \in I\!N)$ an $I\!N$–valued sequence of release variables (to the outside). With a suitable initial value X_0 the sequence*

$$X_{t+1} = X_t - D_t + A_t, \quad t \in I\!N,$$

defines an S–queue if for $0 \leq u \leq v$

$$P(D_t = u | X_s, s \leq t; D_s, s < t; A_s, s \geq 0, X_t + A_t = v) = S(v, u),$$

holds for all $t \in I\!N$. If for all $v \geq 0$

$$S(v, 0) = c(v), \quad \text{and for} \quad u \geq 0$$

$$S(v, u) = \frac{c(v)}{u!} a(v) a(v-1) \cdots a(v - u + 1), \quad 0 < u \leq v,$$

where $a(0) = 1$ and $a(u) > 0$ for $u > 0$, and $c(v)$ is such that $S(v, \cdot)$ is a density on $\{0, 1, \ldots, v\}$, then the queue length process $X = (X_t : t \in I\!N)$ has an invariant measure

$$\hat{\pi}(v) = c \frac{\lambda^v}{a(0) \cdots a(v)}, \quad v \geq 0.$$

If the invariant measure $\hat{\pi}$ can be normalized to an invariant probability law π, then in steady state the departure sequence $D = (D_t : t \in I\!N)$ is an i.i.d. sequence of Poisson–λ variables. Further for any $t \in I\!N$ $(D_s : s \leq t - 1)$ is independent of X_t. (This is the quasi–reversibility for the S–queue.)

Note that the time indices are partially shifted in the model of theorem 6.14 compared to that from that used in [135] to fit it formally into the framework of Section 6.1.

For using quasi–reversibility to prove product form equilibria for networks built from S–queues it turned out that the network has to be observed at time instances when customers are released from the nodes and from the outside but are not yet deposited at the destination nodes, see lemma 12.8 in [18]. Nevertheless, as Henderson and Taylor remarked (see [56], Sect.3.3), these networks fit into the formalism described in Section 6.1.

In fact, these networks with only one customer class are closely related to example 6.10. This is because Walrand assumed independent Poisson–γ_j arrival sequences, $j = 1, \ldots, J$, ($\gamma = \sum_{j=1}^{J} \gamma_j$), Markovian routing, and because the functions $S(\cdot, \cdot)$ can be suitably reproduced. Taking Walrand's routing probabilities $(p(i, j) : 1 \leq i, j \leq J)$ and additionally $(p(0, j) = \gamma_j / \gamma : 1 \leq j \leq J)$ we define

$$\Phi(n) = \prod_{i=1}^{J} c(n_i)^{-1} \prod_{h=1}^{n_i} \alpha_i(h)^{-1}, \quad n \in S,$$

$$\Psi(n - a^+) = \prod_{i=1}^{J} \prod_{h=1}^{n_i - a_i} \alpha_i(h)^{-1}, \quad n \in S, a \in \mathcal{A}, n - a \geq 0,$$

$$\Theta(a) = \frac{e^\gamma \gamma^{a(0)}}{\prod_{i=1}^{J} a(i)!}, \quad a \in \mathcal{A},$$

and obtain for a stable system an equilibrium probability

$$\pi(n) = C\Phi(n) \prod_{i=1}^{J} y(i)^{n_i}, \quad n \in S, \tag{6.18}$$

where C is the normalizing constant and $(y(j) : 1 \leq j \leq J)$ is the solution of the traffic equation similar to (6.16).

The steady state probabilities (6.18) are obtained by Henderson and Taylor ([56]) and differ from that obtained by Walrand. This is due to the fact that the systems are observed at different time points.

Remark 6.15 (General network topology) *While in examples 6.11 and 6.12 quasi reversibility allows to couple nodes to linear tandems, the discussion above shows that S–queues can be used to build up networks with general topology, resp. graph structure. The reason for this is that independent splitting and merging of Poisson variables by the Markovian routing process yields Poisson variables as the batch sizes at other subsequent nodes as well. This is not the case for the batch size distributions in the other systems.*

Walrand's networks of S–queues appear as standard examples in the literature concerning service systems with batch arrivals and batch services. See, e.g, [6], [7], and [8]. Special attention is paid to these networks in the books [142] and [18]. In [142] all networks considered are of this type. For more details and especially for an extensive list of applications we refer to this book.

6.5   Networks with Triggered Batch Movements

Our aim in this section is complement the previous discussion and to sketch further directions of recent reseach concerning more general batch movement networks by describing a specific example.

These networks are generalizations of those described in section 6.1. My description follows Henderson, Northcote, and Taylor [55]. The concept is developed as a unification of the batch movements networks and the recently introduced networks where events (e.g. service completions or arrivals) may trigger other events to happen. This happens in a way, that by a triggering event (e.g.) a sudden departure of specific customers may be enforced without those customers having obtained the full requested service time. A related concept is that of *negative customers*, see [46] and [19]. For a discussion of further related models and for more references see in [18] the reference note 9.8. The book contains generalizations of the network described in section 6.1 in chapters 4 through 11 there. A recent extension of this research was done by Serfozo and Yang [122] who extended the general model description considerably. Continuation of this in continuous time framework to general batch service disciplines is done by Peterson [108]. All these models provide us with discrete time queueing networks as well using the approaches sketched below remark 6.7.

We consider an open network of queues with nodes $1, 2, \ldots, J$. Customers enter the system from the outside (node 0), procede according to some routing regime through the network and eventually leave the system. The customers are undistinguishable and the state description of the network records the joint queue length vectors. Different customer types which may randomly change can be incorporated, see [55] section 4.

The joint queue length process is $X = ((X_t(j) : j = 1, \ldots, J) : t \in I\!N)$, where $X_t(j) = n_j$ indicates that at time t there are n_j customers present at node j. The construction of the network's evolution guarantees that X is a Markov chain with state space $Z\!\!\!Z^J$. The one–step transitions and their conditional probabilities are as follows:

If the state of the network at time $t \in I\!N$ is $X_t = n = (n_1, \ldots, n_J)$, then with probability

$$q(n, a) = \frac{\Psi(n - a)\Theta(a)}{\Phi(n)}$$

a batch $a = (a_1, \ldots, a_J)$ of customers is served at nodes $1, \ldots, J$, respectively.

With probability $p(a, a', a'')$ the released batch a attempts to trigger a batch $a' = (a'_1, \ldots, a'_J)$ (enforcing these customers to immediately finish their service) and to deposit a further batch $a'' = (a''_1, \ldots, a''_J)$ to the respective nodes, $\sum_{a'} \sum_{a''} p(a, a', a'') = 1$. (Here $a, a', a'' \in \mathcal{A}$ have to be chosen assuring some feasibility condition to guarantee consistent transitions.) The external arrivals are included in the deposited batch a'', the internal transitions are included in a, a' and a'', the external departures are due to either a or a'. Batch a' accepts the triggering with probability $\Psi(n - a - a')/\Psi(n - a)$ and rejects it with probability $1 - \Psi(n - a - a')/\Psi(n - a)$.

The state X_{t+1} of the network at time $t + 1$ then is $n - a - a' + a''$ if the triggering was accepted and $n - a$ if it was rejected.

The *traffic equations* similarly defined to (6.8) take the form

$$\Theta(a)f(a) \tag{6.19}$$
$$= \sum_{a' \in \mathcal{A}} \sum_{a'' \in \mathcal{A}} \Theta(a')f(a')[f(a'')p(a', a'', a) - f(a)p(a', a, a'')],$$
$$a \in \mathcal{A} - \{0\}.$$

The introduction of an additional function g similar to Section 6.1 and the required representation (6.9) is substituted here by an *assumption* on the form of an *existing* strict positive solution of (6.19). The main result is:

Theorem 6.16 *If the traffic equation (6.19) has a product solution*

$$f(a) = \prod_{j=1}^{J} y_j^{a_j} > 0$$

then X has an invariant measure

$$\hat{\pi}(n) = \Phi(n)f(n), \quad n \in \mathbb{N}^J.$$

Proof: A direct solution of the global balance equations can be found in [55].

Remark 6.17 *The existence of a strict positive solution of (6.19) is a nontrivial problem. For a discussion see [55], pp132-133.*

Example 6.18 *An example of a telecommunication network where customers may trigger by their arrival additional resources is worked out in detail in [55], Section 4. This network model is then solved using the result of Theorem 6.16 .*

Chapter 7

Appendix

In this appendix we collect for ease of reference some facts which are used in the preceding chapters. Some of these are standard, sometimes not explicitly stated in the literature, some are proved for special requirements in the present text, but their proof is not essential for understanding the main developments.

7.1 Time Reversed Processes

We collect some propositions which are concerned with the time reversal of homogeneous stationary Markov chains. The proofs are similar to those of their continuous time analogues, which are well known. For more details see [75] and [73].

Let $X = (X(t) : t \in \mathbb{Z}), X(t) : (\Omega,\mathcal{F},P) \to (E,\mathcal{P}(E)), t \in \mathbb{Z}$, be a homogeneous ergodic Markov chain with countable state space E, transition probability matrix p. The stationary distribution of X is $\pi = (\pi(x) : x \in E)$.

Proposition 7.1 (Time reversal) *Let X be in equilibrium π and denote by $\overline{X} = (\overline{X}(t) : t \in \mathbb{Z})$ the time reversal of X defined pathwise by*

$$\overline{X}(t,\omega) = X(-t,\omega), \quad t \in \mathbb{Z}, \quad \omega \in \Omega.$$

Then $\overline{X}$ is a homogeneous, stationary ergodic Markov chain with steady state distribution π and transition probability matrix $\bar{p}$ given by

$$\bar{p}(x,y) = p(y,x)\frac{\pi(y)}{\pi(x)}, \quad x,y \in E.$$

Proposition 7.2 (Computing steady states by reversing time) *If for the Markov chain X with transition matrix p and and equilibrium π we can find a transition probability matrix $\tilde{p} = (\tilde{p}(x,y) : x,y \in E)$ such that for some probability measure $\tilde{\pi} = (\tilde{\pi}(x) : x \in E)$*

$$\tilde{\pi}(x)p(x,y) = \tilde{\pi}(y)\tilde{p}(y,x), \quad x,y \in E,$$

H. Daduna: Queueing Networks with Discrete Time Scale, LNCS 2046, pp. 121–123, 2001.
© Springer-Verlag Berlin Heidelberg 2001

holds, then $\pi = \tilde{\pi}$ is the unique steady state of X and $\overline{X}$ and $\tilde{p} = \bar{p}$ is the transition probability matrix of $\overline{X}$.

7.2   Symmetric Functions

Lemma 7.3 *For integers $K, J \geq 1, h \in \{0, 1, \ldots, J\}$, and $\Gamma(h) := \{A \subseteq \{1, \ldots, J\} :| A |= h\}$ we define the function*

$$C_h^{K,J} : \Gamma(h) \times \mathbb{R}^J \longrightarrow \mathbb{R}$$

by

$$(A; z_1, \ldots, z_J) \longrightarrow \sum_{\substack{(x_1, \ldots, x_J) \in \mathbb{N}^J \\ x_1 + \cdots + x_J = K}} \left(\prod_{j \in A} z_j^{x_j} \right) \left(\prod_{j \in \{1, \ldots, J\} - A} z_j^{x_j} \left(1 + \frac{1}{z_j} \right)^{\eta(0, x_j)} \right),$$

$(A, z_1, \ldots, z_J) \in \Gamma(h) \times \mathbb{R}^J$, where $\eta(a, b) = \begin{cases} 0 & a = b \\ 1 & a \neq b \end{cases}$

is the complementary Kronecker delta.
     Then for fixed $(z_1, \ldots, z_J) \in \mathbb{R}^J$ the function

$$C_h^{K,J}(\cdot; z_1, \ldots, z_J) : \Gamma(h) \longrightarrow \mathbb{R}, A \longrightarrow C_h^{K,J}(A; z_1, \ldots, z_J)$$

is a constant. (For definitness we set $0 \cdot \frac{1}{0} := 0$.)

Proof: (a) For $J = 2$ and $h = 1$ $C_1^{K,2}(\{1\}; z_1, z_2) = C_1^{K,2}(\{2\}; z_1, z_2)$ can be verified by direct computations.

(b) Suppose we have proved the statement for all $J = 2, \ldots, J_0 - 1, h \leq J_0 - 1$, and all K. We shall prove the statement for $J_0, h \leq J_0$, and all K by induction.
     Let $A_0, A_1 \in \Gamma(h_0) = \{A \subseteq \{1, \ldots, J_0\} :| A |= h_0\}, A_0 \neq A_1, i_0 \in A_0 - A_1, i_1 \in A_1 - A_0$, and $z_1, \ldots, z_{J_0} \in \mathbb{R} - \{0\}$.
     Then

$$C_{h_0}^{K,J_0}(A_0; z_1, \ldots, z_{J_0})$$

$$= \sum_{k=0}^{K} \left\{ \sum_{\substack{(x_{i_0}, x_{i_1}) \in \mathbb{N}^2 \\ x_{i_0} + x_{i_1} = k}} z_{i_0}^{x_{i_0}} \cdot \left(1 + \frac{1}{z_{i_1}} \right)^{\eta(0, x_{i_0})} \cdot z_{i_1}^{x_{i_1}} \right\}$$

$$\left\{ \sum_{\substack{x_j \in \mathbb{N}, j \in \{1, \ldots, J_0\} - \{i_0, i_1\} \\ \sum_{j \neq i_0, i_1} x_j = K - k}} \left(\prod_{j \in A_0 - \{i_0\}} z_j^{x_j} \right) \right.$$

$$\left. \left(\prod_{j \in \{1, \ldots, J_0\} - \{A_0\} - \{i_1\}} \left(1 + \frac{1}{z_j} \right)^{\eta(0, x_j)} z_j^{x_j} \right) \right\}$$

$$= \sum_{k=0}^{K} \left\{ \sum_{\substack{(x_{i_0}, x_{i_1}) \in I\!N^2 \\ x_{i_0} + x_{i_1} = k}} z_{i_0}^{x_{i_0}} \cdot \left(1 + \frac{1}{z_{i_0}}\right)^{\eta(0, x_{i_0})} \cdot z_{i_1}^{x_{i_1}} \right\}$$

$$\sum_{\substack{x_j \in I\!N, j \in \{1, \ldots, J\} - \{i_0, i_1\} \\ \sum_{j \neq i_0, i_1} x_j = K - k}} \left(\prod_{j \in A_1 - \{i_1\}} z_j^{x_j} \right)$$

$$\left(\prod_{j \in \{1, \ldots, J_0\} - \{A_1\} - \{i_0\}} \left(1 + \frac{1}{z_j}\right)^{\eta(0, x_j)} z_j^{x_j} \right)$$

$$= C_{h_0}^{K, J_0}(A_1; z_1, \ldots, z_{J_0}).$$

Here all substitutions are justified by the induction hypothesis. $\odot$

Corollary 7.4 *For fixed* K, J, h, A *the function*

$$C_h^{K, J}(A; \cdot) : I\!R^J \longrightarrow I\!R, (z_1, \ldots, z_J) \longrightarrow C_h^{K, J}(A; z_1, \ldots, z_J)$$

is a symmetric function, i.e., interchanging the arguments z_j *leaves the value of the function invariant.*

Bibliography

[1] B. van Arem, E. A. van Doorn, and T.M.J. Meijer. Queueing analysis of a discrete closed-loop conveyor with service facilities. *Queueing System*, 4:95–114, 1989.

[2] F. Baccelli and P. Bremaud. *Elements of queueing theory*. Springer, New York, 1994.

[3] A. D. Barbour and R. Schassberger. Insensitive average residence times in generalized semi–markov–processes. *Advances of Applied Probability*, 13:720–735, 1981.

[4] F. Baskett, M. Chandy, R. Muntz, and F.G. Palacios. Open, closed and mixed networks of queues with different classes of customers. *Journal of the Association for Computing Machinery*, 22:248–260, 1975.

[5] G. Bolch, S. Greiner, H. de Meer, and K. S. Trivedi. *Queueing networks and Markov chains*. John Wiley, New York, 1998.

[6] R.J. Boucherie. *Product form in queueing networks*. PhD thesis, Vrije Universiteit Amsterdam, 1992.

[7] R.J. Boucherie and N.M.van Dijk. Spatial birth–death processes with multiple changes and applications to batch service networks and clustering processes. *Advances of Applied Probability*, 22:433–455, 1990.

[8] R.J. Boucherie and N.M.van Dijk. Product forms for queueing networks with state-dependent multiple job transitions. *Advances of Applied Probability*, 23:152–187, 1991.

[9] O. J. Boxma and H. Daduna. Sojourn times in queueing networks. In H. Takagi, editor, *Stochastic Analysis of Computer and Communication Systems*, pages 401–450, Amsterdam, 1990. IFIP, North–Holland.

[10] O. J. Boxma, F. P. Kelly, and A.-G. Konheim. The product form for sojourn time distributions in cyclic exponential queues. *Journal of the Association for Computing Machinery*, 31:128–133, 1984.

[11] O.J. Boxma and J.A.C. Resing. Tandem queues with deterministic service times. *Operations Research*, 49:221–239, 1994.

[12] E. Brockmeyer, H.L. Halstrom, and A. Jensen. *The life and the work of A.K.Erlang*, volume 2 of *Transactions of the Danish Academy of Techn. Science*. Danish Academy of Science, Copenhagen, 1948.

[13] S.C. Bruell and G. Balbo. *Computational algorithms for closed queueing networks*. North–Holland, New York, 1980.

[14] H. Bruneel and Byung G. Kim. *Discrete-Time Models for Communication Systems including ATM*. Kluwer Academic Publications, Boston, 1993.

[15] H. Bruneel, B. Steyaert, E. Desmet, and G.H. Petit. An analytical technique for the derivation of the delay performance of ATM switches with multiverser output queues. *Intern. Journ. of Digital and Analog Communication Systems*, 5:193–201, 1992.

[16] P.J. Burke. The output of a queueing system. *Operations Research*, 4:699–704, 1956.

[17] J.P. Buzen. Computational algorithms for closed queueing networks with exponential servers. *Communications of the ACM*, 16:527–531, 1973.

[18] X. Chao, M. Miyazawa, and M. Pinedo. *Queueing Networks – Customers, Signals, and Product Form Solutions*. Wiley, Chichester, 1999.

[19] X. Chao and M. Pinedo. On generalized networks of queues with positive and negative arrivals. *Prob.Eng.Inf.Sci*, 7:301–334, 1993.

[20] M.L. Chaudhry and U.C. Gupta. Transient behaviour of the discrete time Geom/Geom/m/m Erlang loss model. In A.C. Borthakur and M.L. Choudhry, editors, *Probability Models and Statistics, A.J. Medhi Festschrift*, pages 133 – 145, New Delhi, 1996. New Age International Limited, Publishers.

[21] J. W. Cohen. *The Single Server Queue*. North–Holland Publishing Company, Amsterdam – London, second edition, 1982.

[22] J.W. Cohen. The multiple phase service network with generalized processor sharing. *Acta Informatica*, 12:245–284, 1979.

[23] Coleman, J.L., Henderson, W., Pearce, C.E.M., and Taylor, P.G. A correspondence between product–form batch–movement queueing networks and single–movement networks. *Journal of Applied Probability*, 34:160–175, 1997.

[24] R. B. Cooper. Queueing theory. In D. P. Heyman and M. J. Sobel, editors, *Stochastic Models*, volume 2 of *Handbooks in Operations Research and Management Science*, chapter 10, pages 469– 518. North-Holland, Amsterdam, 1990.

[25] H. Daduna. *Spezielle Stochastische Prozesse*. Institute of Mathematical Stochastics, University of Hamburg, 1984. Lecture notes.

[26] H. Daduna. The cycle time distribution in a cycle of Bernoulli servers in discrete time. *Mathematical Methods of Operations Research*, 44:295 – 332, 1996.

[27] H. Daduna. Stochastische Prozesse. Vorlesungsmanuskript, Institut für Mathematische Stochastik der Universität Hamburg, 1996.

[28] H. Daduna. Discrete time analysis of a state dependent tandem with different customer types. In Christian Freksa, Matthias Jantzen, and Rüdiger Valk, editors, *Foundations of Computer Science, Potential - Theory - Cognition*, volume 1337 of *Lecture Notes in Computer Science*, pages 287–296. Springer, Berlin, 1997.

[29] H. Daduna. The joint distribution of sojourn times for a customer traversing an overtake-free series of queues: The discrete time case. *Queueing Systems and Their Applications*, 27:297–323, 1997.

[30] H. Daduna. Sojourn time distributions in non–product form queueing networks. In Jewgeni Dshalalow, editor, *Frontiers in Queueing: Models and Applications in Science and Engineering*, chapter 7, pages 197–224. CRC Press, Boca Raton, 1997.

[31] H. Daduna. Some results for steady–state and sojourn time distributions in open and closed linear networks of Bernoulli servers with state–dependent service and arrival rates. *Performance Evaluation*, 30:3–18, 1997.

[32] H. Daduna and R. Schassberger. A discrete–time round–robin queue with bernoulli input and general arithmetic service time distributions. *Acta Informatica*, 15:251 –263, 1981.

[33] H. Daduna and R. Schassberger. Networks of queues in discrete time. *Zeitschrift fuer Operations Research ZOR*, 27:159 – 175, 1983.

[34] H. Daduna and R. Schassberger. Delay time distributions and adjusted transfer rates for Jackson networks. *Archiv für Elektronik und Übertragungstechnik*, 47:342 – 348, 1993.

[35] H. Daduna and R. Szekli. Conditional job observer properties in multitype closed queueing networks. Preprint, Mathematical Institute of the University of Wroclaw, 1999.

[36] J. N. Daigle and St. C. Tang. The queue length distribution for multiserver discrete time queues with batch Markovian arrivals. *Comm.Statist.- Stochastic Models*, 8:665–683, 1992.

[37] E. de Souza e Silva and R. R. Muntz. Queueing networks : Solutions and applications. In H. Takagi, editor, *Stochastic Analysis of Computer and Communication Systems*, pages 319–399, Amsterdam, 1990. IFIP, North–Holland.

[38] B. Desert. Lineare stochastische Netzwerke in diskreter Zeit: Gleichge-
wichtsverhalten und Durchlaufzeitverteilungen, 1997. Diploma thesis.

[39] B. Desert and H. Daduna. Discrete time tandem networks of state dependent queues: The effect of different regulation schemes for simultaneous
events on customer oriented performance measures. Preprint 99-07, Institut für Mathematische Stochastik der Universität Hamburg, 1999. sub-
mitted.

[40] N. M. van Dijk. *Queueing Networks and Product Forms – A Systems
Approach.* Wiley, Chichester, 1993.

[41] M. El-Taha and S. Jr. Stidham. A filtered ASTA property. *Queueing
Systems and Their Applications*, 11:211–222, 1992.

[42] M. El-Taha and S. Jr. Stidham. *Sample-Path Analysis of Queueing Systems.* Kluwer Academic Publisher, Boston, 1999.

[43] M. El-Taha, S. Jr. Stidham, and R. Anand. Sample-path insensitivity of
symmetric queues in discrete time. *Nonlinear Analysis, Theory Methods
and Applications*, 30:1099–1110, 1997.

[44] W. Feller. *An Introduction to Probability Theory and Its Applications*,
volume I. John Wiley and Sons, Inc., New York – London – Sidney, third
edition, 1968.

[45] H.-W. Ferng and J.-F. Chang. The departure process of discrete–time
queueing systems with Markovian type input. *Queueing Systems and Their
Applications*, 36:201–220, 2000.

[46] E. Gelenbe. Produkt form queueing networks with negative and positve
customers. *Journal of Applied Probability*, 28:656–663, 1991.

[47] E. Gelenbe and G. Pujolle. *Introduction to queueing networks.* Wiley,
Chichester, 1987.

[48] P. Glasserman. *Gradient estimation via Perturbation Analysis.* Kluwer
Academic Press, Boston, 1991.

[49] B.W Gnedenko and D. König. *Handbuch der Bedienungstheorie*, volume 2.
Akademie–Verlag, Berlin, 1984.

[50] W.J. Gordon and G.F. Newell. Closed queueing networks with exponential
servers. *Operations Research*, 15:252–267, 1967.

[51] A. Gravey and G. Hebuterne. Simultaneity in discrete–time single server
queues with Bernoulli inputs. *Performance Evaluation*, 14:123–131, 1992.

[52] S. Halfin. Batch delays versus customer delays. *The Bell System Technical
Journal*, 62:2011–2015, 1983.

[53] A. Harel, S. Namn, and J. Sturm. Simple bounds for closed queueing networks. *Queueing Systems and Their Applications*, 31:125–135, 1999.

[54] N. Harnpanichpun and G. Pujolle. Product–form discrete–time queues in series with batch transition: late arrival case. *Performance Evaluation*, 21:261–269, 1995.

[55] W. Henderson, B.S. Northcote, and P.G. Taylor. Triggered batch movement in queueing networks. *Queueing Systems and Their Applications*, 21:125 – 141, 1995.

[56] W. Henderson and P.G. Taylor. Product form in queueing networks with batch arrivals and batch services. *Queueing Systems and Their Applications*, 6:71–88, 1990.

[57] W. Henderson and P.G. Taylor. Embedded processes in stochastic Petri nets. *IEEE Transactions on Software Engineering*, 17:108–116, 1991.

[58] W. Henderson and P.G. Taylor. Some new results on queueing networks with batch movements. *Journal of Applied Probability*, 28:409–421, 1991.

[59] W. Henderson and P.G. Taylor. Discrete–time queueing networks with geometric release probabilities. *Advances of Applied Probability*, 24:229–233, 1992.

[60] Henderson, W., Pearce, C.E.M., Taylor, P.G., and Dijk, N.M. van. Closed queueing networks with batch services. *Queueing Systems and Their Applications*, 6:59 – 70, 1990.

[61] Henderson, W., Pearce, C.E.M., Taylor, P.G., and Dijk, N.M. van. Insensitivity in discrete-time generalized semi-Markov processes allowing multiple events and probabilistic service scheduling. *Annals of Applied Probability*, 5:78–96, 1995.

[62] C. Herrmann. *Stochastische Modelle für ATM–Konzepte*. PhD thesis, RWTH Aachen, 1995. Aachener Beiträge zur Mobil– und Telekommunikation.

[63] J. Hofmann, N. Müller, and K. Natarajan. Parallel versus sequential task processing: A new performance model in discrete time. Preprint, Department of Computer Science, University of Trier, 1996.

[64] S.D. Hohl and P.J. Kühn. Approximate analysis of flow and cycle times in queueing networks. In L.F.M. de Moraes, E. de Souza e Silva, and L.F.G. Soares, editors, *Proceedings of the 3rd International Conference on Data Communication Systems and Their Performance*, pages 471–485, Amsterdam, 1988. North–Holland.

[65] J. Hsu and P.J. Burke. Behaviour of tandem buffers with geometric input and markovian output. *IEEE Transactions on Communications*, 24:358 – 361, 1976.

[66] J. J. Hunter. *Mathematical Techniques of Applied Probability*, volume I: *Discrete Time Models: Basic Theory*. Academic Press, New York, 1983.

[67] J. J. Hunter. *Mathematical Techniques of Applied Probability*, volume II: *Discrete Time Models: Techniques and Applications*. Academic Press, New York, 1983.

[68] F. Ishizaki and T. Takine. Loss probability in a finite discrete–time queue in terms of the steady state distribution of an infinite queue. *Queueing Systems and Their Applications*, 31:317 –326, 1999.

[69] J.R. Jackson. Networks of waiting lines. *Operations Research*, 5:518–521, 1957.

[70] S. Karlin and H. M. Taylor. *A First Course in Stochastic Processes*. Academic Press, New York – San Francisco – London, second edition, 1975.

[71] S. Karlin and H. M. Taylor. *A Second Course in Stochastic Processes*. Academic Press, New York – San Francisco – London, 1981.

[72] U. S. Karmarkar. Manufacturing lead times, order release and capacity loading. In S.C. Graves, A.H.G. Rinnooy Kan, and P.H. Zipkin, editors, *Logistics of Production and Inventory*, volume 4 of *Handbooks in Operations Research and Management Science*, chapter 6, pages 287–329. North–Holland, Amsterdam, 1993.

[73] J. Keilson. *Markov chain models – Rarity and exponentiality*. Springer, New York, 1979.

[74] F. Kelly. Networks of queues. *Advances of Applied Probability*, 8:416–432, 1976.

[75] F. P. Kelly. *Reversibility and Stochastic Networks*. John Wiley and Sons, Chichester – New York – Brisbane – Toronto, 1979.

[76] F. P. Kelly and Phillip Pollett. Sojourn times in closed queueing networks. *Advances of Applied Probability*, 15:638–653, 1983.

[77] J. G. Kemeny, J. L. Snell, and A. W. Knapp. *Denumerable Markov Chaines*. Springer–Verlag, New York – Heidelberg – Berlin, 1976. Reprint of the book published in 1966 by Van Nostrand, Princeton.

[78] L. Kleinrock. Analysis of a time–shared processor. *Naval Research Logistics Quarterly*, 10(II):59–73, 1964.

[79] L. Kleinrock. *Communication nets – Stochastic message flow and delay*. McGraw–Hill, New York, 1964.

[80] L. Kleinrock. Time–shared systems: A theoretical treatment. *Journal of the Association for Computing Machinery*, 14(2):242–261, 1967.

[81] L. Kleinrock. *Queueing Theory*, volume I. John Wiley and Sons, New York, 1975.

[82] L. Kleinrock. *Queueing Theory*, volume II. John Wiley and Sons, New York, 1976.

[83] H. Kobayashi. Stochastic modeling: Queueing networks. In G. Louchard and G. Latouche, editors, *Probability Theory and Computer Science*, International Lecture Series in Computer Science, chapter Part II, pages 53–121. Academic Press, London,Orlando, 1983.

[84] D. Koenig and V. Schmidt. EPSTA: The coincidence of time-stationary and customer-stationary distributions. *Queueing Systems and Their Applications*, 5:247–264, 1989.

[85] E. Koenigsberg. Twenty five years of cyclic queues and closed queue networks: A review. *Journal of the Operational Research Society*, 33:605–619, 1982.

[86] K.H. Kook and R.F. Serfozo. Travel and sojourn times in stochastic networks. *Annals of Applied Probability*, 3:228–252, 1993.

[87] D. Kouvatsos. *Performance Evaluation and Applications of ATM Networks*, volume 557 of *The Kluwer International Series in Engineering and Computer Science*. Kluwer, Boston, 2000.

[88] D. D. Kouvatsos, N. M. Tabet-Aouel, and S. G. Denazis. ME–based approximations for general discrete–time queueing models. *Performance Evaluation*, 21:81–109, 1994.

[89] M. Krüger. Geschlossene Warteschlangen–Netzwerke unter doppelt–stochastischen Bediendisziplinen. Diplomarbeit Technische Universität Berlin, Fachbereich Mathematik, 1983.

[90] P. R. Kumar. Re–entrant lines. *Queueing Systems and Their Applications*, 13:87–110, 1993.

[91] P. R. Kumar. Scheduling manufacturing systems of re–entrant lines. In D. D. Yao, editor, *Stochastic Modeling and Analysis of Manufacturing Systems*, Springer Series in Operations Research, chapter 8, pages 325–360. Springer, New York, 1994.

[92] K. Laevens. The round–robin service discipline in discrete time for phase–type distributed packet–lengths. Preprint, SMAC Research Group, University of Ghent, 1996.

[93] K. Laevens and H. Bruneel. Discrete–time queueing models with feedback for input–buffered ATM switches. *Performance Evaluation*, 27,28:71–87, 1996.

[94] S. S. Lavenberg and M. Reiser. Stationary state probabilities at arrival instants for closed queueing networks with multiple types of customers. *Journal of Applied Probability*, 17:1048–1061, 1980.

[95] Louchard, G. and Latouche, G., editors. *Probability theory and computer science*. International Lecture Series in Computer Science. Academic Presss, New York, 1983.

[96] A. Makowski, B. Melamed, and W. Whitt. On averages seen by arrivals in discrete time. In *IEEE Conference on Decision and Control, Vol. 28*, pages 1084–1086, Tampa, FL., 1989.

[97] Benjamin Melamed and Wy Whitt. On arrivals that see time averages. *Operations Research*, 38:156–172, 1990.

[98] M. Miyazawa. On the characterisation of departure rules for discrete–time queueing networks with batch movements and its applications. *Queueing Systems and Their Applications*, 18:149–166, 1994.

[99] M. Miyazawa. A note on my paper: On the characterisation of departure rules for discrete–time queueing networks with batch movements and its applications. *Queueing Systems and Their Applications*, 19:445–448, 1995.

[100] M. Miyazawa. Stability of discrete–time Jackson networks with batch movements. In Paul Glasserman, Karl Sigman, and David D. Yao, editors, *Stochastic Networks: Stability and Rare Events*, volume 117 of *Lecture Notes in Statistics*, chapter 5, pages 76–93. Springer, New York, 1996.

[101] M. Miyazawa and H. Takagi. Editorial introduction to: Advances in discrete time queues, (Special issue of Queueing Systems ,Theory and Applications). *Queueing Systems and Their Applications*, 18:1–3, 1994.

[102] M. Miyazawa and Y. Takahashi. Rate conservation principle for discrete–time queues. *Queueing Systems and Their Applications*, 12:215–230, 1992.

[103] J. A. Morrison. Two discrete time queues in tandem. *IEEE Transactions on Communications*, 27(3):563–573, 1979.

[104] H. Osawa. Quasi–reversibility of a discrete–time queue and related models. *Queueing Systems and Their Applications*, 18:133–148, 1994.

[105] H. G. Perros. Approximation algorithms for open queueing networks with blocking. In H. Takagi, editor, *Stochastic Analysis of Computer and Communication Systems*, pages 451–498. North-Holland, Amsterdam, 1990.

[106] V. Pestien and S. Ramakrishnan. Asymptotic behavior of large discrete–time cyclic queueing networks. *Annals of Applied Probability*, 4:591 – 606, 1994.

[107] V. Pestien and S. Ramakrishnan. Features of some discrete–time cyclic queueing networks. *Queueing Systems and Their Applications*, 18:117 – 132, 1994.

[108] S. Peterson. General batch service disciplines - A product–form batch processing network with customer coalescence. *Mathematical Methods of Operations Research*, 52:79–97, 2000.

[109] G. Pujolle, J.P. Claude, and D. Seret. A discrete queueing system with a product form solution. In Hasegawa, T., Takagi, H., and Takahashi, Y., editors, *Proceedings of the IFIP WG 7.3 International Seminar on Computer Networking and Performance Evaluation*, pages 139 –147, Amsterdam, 1986. Elsevier Science Publisher.

[110] M. Reiser. A queueing network analysis of computer communication networks with window flow control. *IEEE Transactions in Communications*, COM-27:1199–1209, 1979.

[111] M. Reiser. Performance evaluation of data communication systems. *Proceedings of the IEEE*, 70:171–196, 1982.

[112] J.-F. Ren, J. W. Mark, and J.W. Wong. Performance analysis of a leaky-bucket controlled ATM multiplexer. *Performance Evaluation*, 19:73–101, 1994.

[113] Y. Sakai, Y. Takahashi, and T. Hasegawa. Discrete time multi–class feedback queue with vacations and close time under random order of service discipline. *Journal of the Operartions Research Society of Japan*, 41:589–609, 1998.

[114] M. Sakata, S. Noguchi, and J. Oizumi. An analysis of the M/G/1 queue under round–robin scheduling. *Operations Research*, 19:371–385, 1971.

[115] R. Schassberger. Insensitivity of steady–state distributions of generalized semi–Markov processes, part I. *Ann. Prob.*, 5:87–99, 1977.

[116] R. Schassberger. Insensitivity of steady–state distributions of generalized semi–Markov processes, part II. *Ann. Prob.*, 6:85–93, 1978.

[117] R. Schassberger. The doubly stochastic server: A time–sharing model. *Zeitschrift fuer Operations Research ZOR*, 25:179–189, 1981.

[118] R. Schassberger and H. Daduna. A discrete–time technique for solving closed queueing network models of computer systems. In Paul J. Kühn and K.M. Schulz, editors, *Messung,Modellierung und Bewertung von Rechensystemen*, pages 122–134, Berlin, 1983. Informatik–Fachberichte 61, Springer.

[119] R. Schassberger and H. Daduna. The time for a roundtrip in a cycle of exponential queues. *Journal of the Association for Computing Machinery*, 30:146–150, 1983.

[120] R. F. Serfozo. Queueing networks with dependent nodes and concurrent movements. *Queueing Systems and Their Applications*, 13:143–182, 1993.

[121] R. F. Serfozo. *Introduction to Stochastic Networks*, volume 44 of *Applications of Mathematics*. Springer, New York, 1999.

[122] R. F. Serfozo and B. Yang. Markov network processes with string transitions. *Annals of Applied Probability*, 8:793–821, 1998.

[123] K. C. Sevcik and I. Mitrani. The distribution of queueing network states at input and output instants. *Journal of the Association for Computing Machinery*, 28:358–371, 1981.

[124] V. Sharma. Open queueing networks in discrete time – some limit theorems. *Queueing Systems and Their Applications*, 14:159–175, 1993.

[125] V. Sharma and N. D. Gangadhar. Some algorithms for discrete time queues with finite capacity. *Queueing Systems and Their Applications*, 25:281–305, 1997.

[126] K. Sohraby and J. Zhang. Spectral decomposition approach for transient analysis of multi–server discrete–time queues. *Performance Evaluation*, 21:131–150, 1994.

[127] H. Takagi. *Stochastic Analysis of Computer and Communication Systems*. North–Holland, Amsterdam, 1990.

[128] H. Takagi. *Queueing Analysis: A Foundation of Performance Analysis*, volume 3. North–Holland, New York, 1993. Discrete-Time Systems.

[129] H. Tempelmeier. Inventory control using a service constraint on the expected customer order waiting time. *European Journal of Operational Research*, 19:313–323, 1983.

[130] H. Tempelmeier. Inventory service-levels in the customer supply chain. *Operations Research Spektrum*, 22:361–380, 2000.

[131] D. D. Tjhie. *Zeitkritischer Verkehr in Wartesystemen von Hochgeschwindigkeitsnetzen : Modellbildung und Mathematische Analyse*. Herbert Utz Verlag Wissenschaft, M"unchen, 1996.

[132] P. Tran-Gia. *Analytische Leistungsbewertung verteilter Systeme*. Springer, Berlin, 1996.

[133] P. Tran-Gia, C. Blondia, and D. Towsley. Editorial introduction to: Discrete–time models and analysis methods, (Special issue of Performance Evaluation). *Performance Evaluation*, 21:1–2, 1994.

[134] P. Tran-Gia and R. Dittmann. A discrete–time analysis of the cyclic reservation multiple access protocol. *Performance Evaluation*, 16:185–200, 1992.

[135] J. Walrand. A discrete–time queueing network. *Journal of Applied Probability*, 20:903 – 909, 1983.

[136] J. Walrand. *An introduction to queueing networks*. Prentice –Hall International Editions, Englewood Cliffs, 1988.

[137] J. Walrand. Queueing networks. In D. P. Heyman and M. J. Sobel, editors, *Stochastic Models*, volume 2 of *Handbooks in Operations Research and Management Science*, chapter 11, pages 519–603. North–Holland, Amsterdam, 1990.

[138] W. Whitt. An overview of Brownian and non–Browninan FCLTs for the single–server queue. *Queueing Systems and Their Applications*, 36:39–70, 2000.

[139] P. Whittle. *Systems in Stochastic Equilibrium*. Wiley, Chichester, 1986.

[140] S. Wittevrongel and H. Bruneel. Discrete-time ATM queues with independent and correlated arrival streams. In D. D. Kouvatsos, editor, *Performance evaluation and applications of ATMnetworks*, volume 557 of *The Kluwer international series in engineering and computer science*, chapter 16, pages 387–412. Kluwer, Boston, 2000.

[141] R.W. Wolff. Poisson arrivals see time averages. *Operations Research*, 30:223–231, 1982.

[142] M.E. Woodward. *Communication and Computer Networks: Modelling with Discrete–Time Queues*. IEEE Computer Society Press, Los Alamitos, CA, 1994.

[143] S.F. Yashkov. Properties of invariance of probabilistic models of adaptive scheduling in shared–use systems. *Automatic control and computer science*, 14:46–51, 1980.

Index

Lecture Notes in Computer Science

For information about Vols. 1–2025
please contact your bookseller or Springer-Verlag